Microbiological Aspects of

BIOFILMS and DRINKING WATER

The Microbiology of
EXTREME AND UNUSUAL ENVIRONMENTS

SERIES EDITOR
RUSSELL H. VREELAND

Titles in the Series

The Biology of Halophilic Bacteria
Russell H. Vreeland and Lawrence Hochstein

The Microbiology of Deep-Sea Hydrothermal Vents
David M. Karl

The Microbiology of Solid Waste
Anna C. Palmisano

The Microbiology of the Terrestrial Subsurface
Penny S. Amy and Dana L. Haldeman

The Microbiology and Biogeochemistry of Hypersaline Environments
Aharon Oren

Microbiological Aspects of Biofilms and Drinking Water
Steven L. Percival, James T. Walker, and Paul R. Hunter

Microbiological Aspects of BIOFILMS and DRINKING WATER

STEVEN L. PERCIVAL
JAMES T. WALKER
PAUL R. HUNTER

CRC Press
Boca Raton London New York Washington, D.C.

Library of Congress Cataloging-in-Publication Data

Percival, Steven L.
Microbiological Aspects of Biofilms and Drinking Water / Steven L. Percival, James T. Walker, Paul R. Hunter.
p. cm. -- (The microbiology of extreme and unusual environments)
Includes bibliographical references and index.
ISBN 0-8493-0590-X (alk. paper)
1. Biofilms. 2. Drinking water--Microbiology. I. Walker, James Thomas. II. Hunter, Paul R. III. Title. IV. Series.
QR100.8.B55 P47 2000
628.1'6--dc21 99-098186
CIP

Direct all inquiries to CRC Press LLC, 2000 N.W. Corporate Blvd., Boca Raton, Florida 33431.

International Standard Book Number 0-8493-0590-X
Library of Congress Card Number 99-098186
Printed in the United States of America 1 2 3 4 5 6 7 8 9 0
Printed on acid-free paper

About the Authors

Dr. Steven Percival is qualified from the University of Leeds with a PhD in microbiology specilising in biofilmology. His PhD involved looking at the development and consequences of biofilms in drinking water. He also has an MSc in public health and various other qualifications in microbiology. Presently, Dr. Percival is a senior lecturer in microbiology at University College Chester and head of the Microbiology Research Group. Steven has a broad range of experiences in the problems, detection, and control of biofilms in the water and medical industries and has gained a large amount of experience in microbiology and waterborne diseases as a result of this. His research at present involves looking at *Helicobacter pylori*, *Aeromonas hydrophila,* and *Mycobacterium avium intracellulare* complex in biofilms and drinking water systems. Other projects he is involved in include antibiotic and biocidal resistance, the effects of heavy metals on the attachment of pathogens to surfaces, biodeterioration, and microbially induced corrosion. He has been involved in microbiological consultancy for a number of companies, in particular the Ministry of Defence. Dr. Percival is a member of various organisations including the Society for Microbiology, Society for Applied Microbiology, The International Biodeterioration Society, Biofilm Club, and the International Water Association. He is the secretary of the International Biodeterioration Society and microbiological advisor to the membership committee of the Institute of Biology.

Dr. Jimmy Walker obtained an HND in biology at Bellshill College, Scotland before graduating in microbiology from the University of Aberdeen. Jimmy undertook his PhD whilst working at CAMR, investigating biofilms in copper tube corrosion and the survival of *Legionella pneumophila*.

Jimmy has a broad range of experience in the problems, detection, and control of biofilms in the water and medical industries. As a research microbiologist at CAMR, Dr. Walker carries out projects on biofouling often involving category III pathogens such as *E. Coli* 0157. He is an editorial board member of the *International Biodeterioration & Biodegradation Journal* and *Anti-Corrosion Methods & Materials Journal*, and he has published over 40 scientific papers.

Dr. Walker is an external PhD examiner at the Robert Gordon's University in Aberdeen and an external supervisor at the University College Chester. As well as sitting on the committee of the Biofilm Club, he is also the vice president of the International Biodeterioration Society.

Prof. Paul Hunter qualified in medicine from Manchester University and then went on to specialise in medical microbiology. He gained his MD for research into the epidemiology of Candida infection. Prof. Hunter is a fellow of the Royal College of Pathologists and a member of the Faculty of Public Health of the Royal College of Physicians. He was appointed director of the Chester Public Health Laboratory in 1988 and works as a consultant in medical microbiology, communicable disease control, and epidemiology. He is also visiting professor in microbiology at University College Chester. He has had a continuing interest in water microbiology and waterborne disease for many years and has written *Waterborne Disease: Epidemiology and Ecology* and he has published over 100 papers in the scientific and medical literature. Prof. Hunter is chair of the PHLS Advisory Committee on Water and the Environment and serves on several other national and international committees and advisory groups.

Contents

Preface

After many years of studying microbiology, biofilms, and public health, it became my ambition to produce a book on these three areas. Having researched substantially into the formation and development of biofilms, particularly in potable water, I felt a book that could consolidate all the information on their public health significance was needed. This book provides a snapshot of public health and water with an appreciation of what a biofilm is and how well it presents a safe haven for pathogens, while factorizing unreported and reported water-related diseases.

This book has been written with the help of friends, Dr. Jimmy Walker and Professor Paul Hunter, without whom areas of the book would have been difficult to write. The book is written with a large number of people in mind, but in particular, students, lecturers, researchers, and practitioners in water-related problems.

This text is an overview of the public health effects associated with potable water and includes particular reference to the microbiological aspects relating to the development of biofilms. The first five chapters focus on the state of the water supply of the nation, highlighting historical developments and areas of concern. Methods that could be employed to study the epidemiological spread of waterborne infections and methods which are used in surveillance and control of pathogenic microbes are reviewed. Also included is a chapter on legislation and methods which are presently employed for the detection of indicator microorganisms of public health importance in potable water.

Chapters 6 to 11 focus particularly on biofilm development within potable water, highlighting the public health threat from this. Also included here is a very large overall review of the microbes of public health importance in potable water and biofilms. Methods used to detect biofilms can be found in Chapter 9. This by no means includes all the methods that can be used to study biofilms but rather it incorporates a large number of methods which have been shown to help in the analysis. Control of biofilms and the methods that are presently involved, including both conventional and biocidal treatments, are reviewed in the final chapter. The particular control methods covered include chlorination, and the modes of action of this and other biocides are also documented. We hope you enjoy reading this book which will hopefully provide an aid to the study of biofilms in drinking water.

1 Water Supply, Treatment, and Distribution

CONTENTS

1.1 WATER SUPPLY

The ready availability of potable water is taken for granted by most people in the Western world. Nevertheless, as will be seen throughout this book, the need for effective water extraction, treatment, and distribution remains as great now as it ever was. To safeguard public health, potable water has to be free of pathogens and noxious chemicals. It must also be pleasant to taste and have good appearance for human consumption. Water companies have a responsibility to provide a continuous supply of wholesome water which is achievable through the collection, treatment, and, then, distribution of water.

It was, however, not that long ago that safe clean water was not something that people in industrial cities could expect. Even in today's world, the problem of supplying safe water to many of the world's population seems to be becoming beyond our ability to solve. By 2025, one third of the world's population is expected to suffer from chronic water shortages and more than 60% is likely to face some water-related problems. It has also been estimated that the per capita supply of fresh, safe water for the next generation will be only one third of the supply available 30 years ago. Although physiologically man can survive on just a few litres per day, the World Health Organization has suggested that people need 20 to 40 litres per day.[1] The average daily consumption per person in the U.K. is 150 litres, and in the U.S. it ranges from 380 to 950 litres per person per day.[2]

In this chapter we shall briefly review the history of water treatment and distribution technology before discussing in more detail current water treatment technology.

1.2 A SHORT HISTORY OF WATER SUPPLY AND TREATMENT

Although man has used irrigation for agricultural purposes since prehistoric times, it was not until ancient Egyptian and Babylonian civilisations that large-scale irrigation systems of dams and canals were developed.[2] Although their primary purpose was agricultural, individuals undoubtedly used them for their own supply of drinking water. The first civilisation to practice water treatment on a large scale was the Romans. In addition to constructing huge aqueducts to carry clean water from the mountains into the city, they also built settling basins and filters to improve the clarity of the water. With the decline of the Roman Empire, people largely reverted to local wells, springs, and streams.

The first pump-powered water distribution system was built in 1562 in London. This pumped water from the Thames into a reservoir which then distributed the water locally through lead pipes. No particular treatment was applied.

The main impetus for the next stage in the development of water treatment was the cholera and typhoid epidemics which ravaged Europe during the 19th century. The high mortality from these epidemics drove the sanitary movement in the U.K. which put forward the conviction that health and disease were functions of social conditions. This movement led to the 1842 publication of the *Report of an Inquiry into the Sanitary Condition of the Labouring Population of Great Britain* by Edwin Chadwick. This report proposed that ill health was owing to overcrowding, inadequate waste disposal, polluted water, and bad diet. Many would date modern public health to the publication of this report.

It was also around this time when the germ theory of disease started to gain ground. Ironically, many within the sanitary movement had serious objections to the germ theory of disease. We now know that much of what they were recommending worked because of the impact of their suggestions for healthy living on the transmission of microbial pathogens.

Further support for the importance of clean water came from John Snow (1813–1858), who wrote his account, *On the Mode of Communication of Cholera* in 1849. This suggested a correlation between cholera and water supplies. In 1854, Snow was able to demonstrate the link between the Broad Street pump and many fatal cases of cholera. In doing so, he simultaneously proved that water could carry disease and established the science of epidemiology.

In 1846, the U.K. Parliament passed the Liverpool Sanitary Act which gave the city council the power to appoint a medical officer, borough engineer, and inspector of nuisances. Two years later, in 1848, The National Health Act was passed. This act created a general board of health and allowed for the setting up of local boards of health to deal with various matters of environmental cleanliness including water supplies. This act was later replaced by the Public Health Act in 1875.[3] It was from this time that developments in water treatment progressed apace.

1.3 WATER TREATMENT SOURCES

In order to provide a continuous supply of potable water, the most important factor is access to source water of an acceptable quality. In the U.K., the supply of water is generally obtained from surface water which accounts for 70% of available water, and groundwater which accounts for 30% of available water. Whilst surface water is generally easier to both locate and extract than groundwater, it is readily affected by the surrounding environment resulting in the need for greater treatment.[4] Extraction and treatment requirements are very different for surface and groundwater and they will be considered separately.

1.3.1 Surface Water

Surface water includes water from lakes, ponds, rivers, and streams. Ultimately, all surface water falls as rain. Although some of this rain falls directly onto the water body, most falls onto land and then gains access to the water body as runoff. In many lowland waters, much of the water body is filled from other sources such as wastewater treatment plant discharge.

The quality of surface water in upland areas differs substantially from that in lowland areas. Lowland waters are usually more turbid, more nutrient rich, and contain various natural and man-made pollutants. The quality of surface water can also show large temporal variations in water quality owing to factors such as water flow, local rainfall, temperature, and changing industrial activity.

Where rain falls onto impervious rocks, it will flow over the ground or through the soil layer directly into streams or rivers. Such water will have a relatively low mineral content and be quite soft. It can, however, pick up organic and inorganic contaminants. This is most obvious in water from peaty areas which can be highly coloured. If water falls onto porous rock such as chalk or limestone, much of it may soak into the ground. Such water may remain as groundwater, or it may return to the surface as springwater, or directly enter a river from ground flow. Surface water in these more porous areas is more mineralised and harder, but it generally has fewer contaminants and is clearer. Figure 1.1 shows the various routes through which rainwater can enter a water body.

Upland water usually requires very little treatment. There are many communities that take such water abstracted from naturally formed or dammed lakes. Typically, water treatment may include some form of coarse screening with or without filtration and little or no chlorination before being distributed to consumers' homes. The main risk to such supplies comes from agricultural activity in the river's watershed. With chlorinated supplies, the main risk is parasitic disease such as giardiasis and cryptosporidiosis. With un-chlorinated supplies there is also a risk of bacterial pathogens, most commonly, *Campylobacter*.[5] If there is much human habitation above the water extraction point, the range of potential pathogens increases considerably. Control of this risk relies on managing human and agricultural activity in the watershed.

As already mentioned, lowland surface water is usually much more polluted than upland water and so requires considerably more treatment. Sources of pollution for lowland waters include discharges from wastewater treatment plants, industrial sites, and runoff from urban areas as well as agricultural land. Because lowland

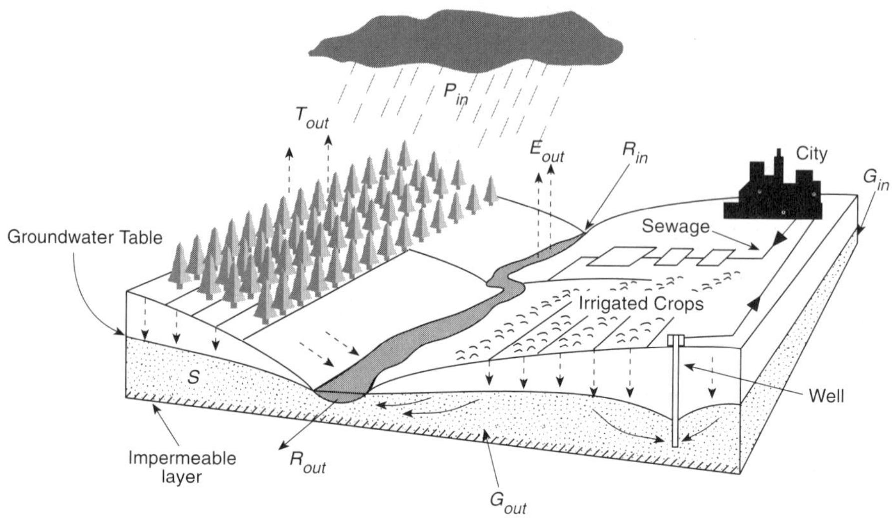

FIGURE 1.1 Routes of water entry into a water body. P = precipitation; T = transpiration; E = evaporation; R = river water; G = ground water; and S = saturated zone. (Courtesy of Davis, M.L. and Cornwell, D.A., *Introduction to Environmental Engineering*, copyright 1991, McGraw-Hill. Material is reproduced with permission of The McGraw-Hill Companies.)

agriculture tends to be more intensive than upland agriculture, lowland agricultural runoff poses a greater problem for water treatment owing to the presence of agrochemicals and faecal contamination. Because of these various sources of pollution, the microbiological and chemical quality of lowland water is variable and can often be very poor. Such water requires considerably more treatment than upland water. Commonly, it is first pumped to a pretreatment reservoir and then subject to several of the various water treatments discussed later on.

1.3.2 Groundwater

Groundwater is that water which is present in soils and rocks which are fully saturated. Groundwater represents by far the largest source of fresh water other than that contained in ice sheets. Most groundwater is ultimately derived from rain which fell onto porous rocks or flowed into rivers or other water bodies which overlay porous rock. Sometimes, the time that water has resided in the ground since the time it fell as rain runs into thousands of years.

Those rock formations that are sufficiently saturated by water to allow extraction are called aquifers. Underlying aquifers are less porous rocks known as aquitards. When an aquitard overlies water bearing rock, it is known as a confined aquifer. In certain circumstances, pressure can build up in a confined aquifer such that an artesian well (one where the water has sufficient pressure to reach the surface without pumping) can be drilled. In an unconfined aquifer, the upper surface of the saturated layer is called the water table. Groundwater can be extracted through wells and boreholes or it can be collected as it naturally leaves the ground as a spring (see Figure 1.2).

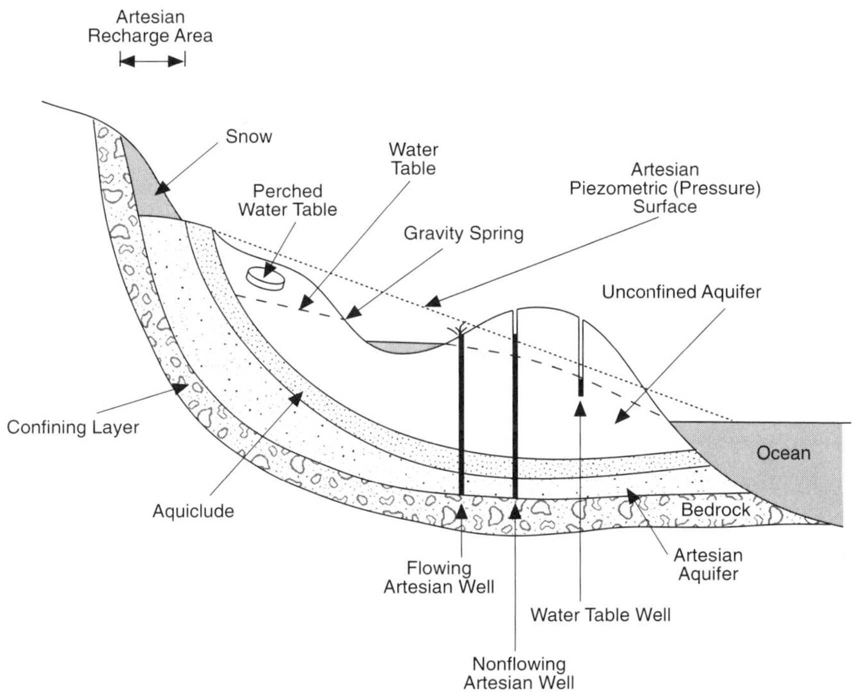

FIGURE 1.2 Groundwater hydrology. (Courtesy of Davis, M.L. and Cornwell, D.A., *Introduction to Environmental Engineering,* copyright 1991, McGraw-Hill. Material is reproduced with permission of The McGraw-Hill Companies.)

Generally the chemical and microbiological composition of groundwater remains fairly constant from one season to the next. This does depend, however, on the recharge time. Springs which take water with very short recharge times do demonstrate significant variations in their character. The nature of the groundwater also varies considerably from site to site depending on such factors as the residence time, nature of the rock, and source of pollution. Aquifers in limestone are much harder than those in sandstone. Also, those that have a high residence time are likely to be more mineralised than those with a low residence time. On the other hand, aquifers with a low residence time are more likely to contain higher organic matter and be more exposed to microbial and chemical contamination.

Because groundwater has been through a filtering process owing to its passage through rock, it is usually less turbid and free from faecal contamination and, therefore, requires less treatment.[6] However, it is a mistake to assume that groundwater is always free from such faecal pathogens. Many waterborne outbreaks have occurred because water suppliers assumed that groundwater was not at risk from faecal pollution.[5]

Pedley and Howard [7] have discussed the various potential sources of groundwater contamination. A major source of contamination is inadequate protection of abstraction points. This will occur if the well is inadequately sealed, animals are allowed to graze, or pit latrines are located too close to the well. Contamination of the aquifer may occur at the point of recharge when wastewater and sewage are spread onto land or allowed into rivers which are contiguous with aquifers. Finally, aquifers may become contaminated from leaking septic tanks, cesspools, sewers, and landfill sites.

1.4 WATER TREATMENT

For the remainder of this chapter we shall discuss the various treatment processes which our water is subject to before it is delivered to the tap. In order to illustrate all the processes involved, we shall consider the various processes which water from a moderately contaminated river may require. Groundwater will usually require much less treatment, which in some cases may mean simply, chlorination. Figure 1.3 illustrates the various processes that may be applied.

1.4.1 PRETREATMENT

Water taken from rivers will need to pass through coarse intake screens to prevent large floating objects from entering the treatment works. Such objects may range from dead leaves to dead sheep. The intake screens are usually made of fairly substantial steel bars.

After intake, water is frequently stored for some time in a pretreatment storage reservoir. Such reservoirs serve several purposes. They provide a useful buffer to even out flow into the treatment works, for example, when the water source is a flashy stream. They also enable the treatment works to stop extraction from the river if there is a marked drop in quality which may follow a pollution incident upstream of the extraction point. Storage also improves water quality before further management. For poor quality waters, such improvement is essential before further treatment. During storage, a considerable amount of particulate matter will settle out of the water body. Also, the impact of sunlight on the water will kill many bacteria and bleach out colour and many chemicals that could cause bad taste. Therefore, during storage total colony and coliform counts will decline, colour will reduce, turbidity will fall, and the water will become better oxygenated. On the other hand, there is the risk that cyanobacteria may bloom and, if the reservoir is not protected, faecal pollution from animals may occur.

From the storage reservoir, water usually passes through fine screens before carried to the rest of the treatment works. For most reasonably sized treatment works, these screens will be mechanical and based on a drum or band principle so that they do not become clogged by silt. For low quality source water, prechlorination may be used further to improve quality. Chlorine may be added soon after extraction from the river. As well as being effective in reducing bacterial counts, chlorine oxidises and precipitates iron and manganese and reduces colour. Unfortunately, rather high levels of chlorine need to be added, owing to the high chlorine demand of raw water. When the water is very turbid, prechlorination is not effective.

Aeration improves the quality of certain waters in several ways. It will improve the amount of dissolved oxygen in water. The treatment of deoxygenated water is often suboptimal. In addition, aeration can improve taste by precipitating out hydrogen

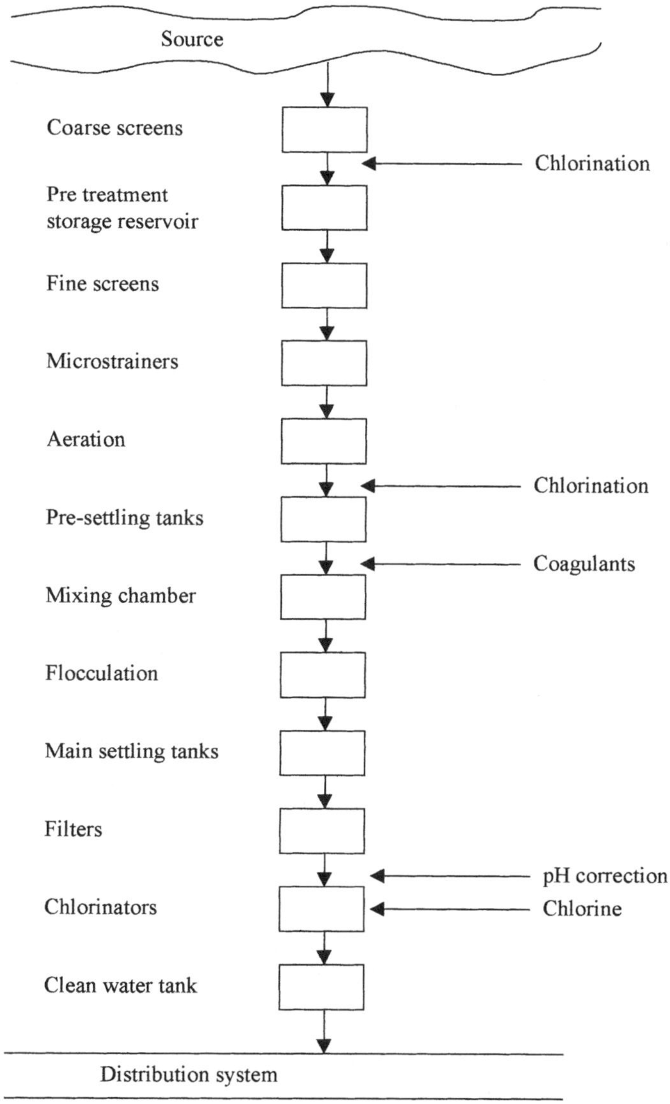

FIGURE 1.3 Flow diagram showing possible treatment stages (adapted from Smethurst[26]).

sulphide or encouraging the release of volatile chemicals such as those produced by algal growth. Carbon dioxide, which can make water corrosive to concrete, is also removed. Finally, excess iron or manganese may also be precipitated out by aeration. Although water can be aerated by passing bubbles up through the water body, usually some form of cascade system is used. Yet another type of aerator works by spraying jets of water at a metal plate.

1.4.2 Coagulation and Sedimentation

After pretreatment, most waters still carry a significant burden of small particulate matter (less than 10 mm) which will not settle out by itself. Such particulates include bacteria and viruses, and organic and inorganic particulates. The function of the next stage in treatment is to remove these particles. In order to achieve this, small particles have to be encouraged to group together into larger particles by the addition of a chemical coagulant. This coagulant adds positive ions to the water to reduce the effect of the negative charges on the surface of colloids in the water to the point that they are no longer repelled by each other. Once this has been achieved, such particles will more easily sediment out or be removed by filtration. The three key properties of coagulants are that they are trivalent cations, nontoxic, and insoluble at a neutral pH.

Traditionally, coagulants were salts of iron or aluminium, though other more modern coagulant chemicals are now being used. The most common coagulants include aluminium sulphate, sodium aluminate, ferrous sulphate, ferric chloride, and ferric sulphate. The coagulants react chemically with alkali in the water to produce gelatinous precipitates (floc). For example, the main chemical reaction for aluminium in a normal water is

$$Al_2(SO_4)_3 + 3Ca(HCO_3) \rightarrow 2Al(OH)_3 + 3CaSO_4 + 6CO_2$$

If lime has been added this becomes

$$Al_2(SO_4)_3 + 3Ca(OH)_2 \rightarrow 2Al(OH)_3 + 3CaSO_4$$

In both cases, the aluminium hydroxide is insoluble and precipitates out.

It may be necessary to add a coagulant aid as well as a coagulant. pH adjusters such as sulphuric acid or lime may be added to adjust the pH to that which is optimal for coagulation. Activated silica has a negative surface charge and can combine with coagulant to produce a larger denser floc even at low doses. Sometimes, clays can be used in a similar way to activated silica. Finally, polyelectrolytes may also be used. Polyelectrolytes are synthetic, long chain carbon polymers with variable surface charge and several active sites which bind to flocs, producing larger, tougher flocs by a process known as interparticle bridging. However, the dose of polyelectrolyte needs to be controlled very carefully.

The coagulant must be added very rapidly for maximum efficiency. A variety of systems have been devised to aid rapid mixing. Whilst floc formation is almost instantaneous once the coagulant has been added, the flocs are quite small. During the next step of flocculation, these small flocs coalesce with others to produce larger ones. This is aided by gentle mixing. If mixing is too violent, these flocs break up. Many of these larger flocs are then allowed to sediment out of the water in a sedimentation basin, also known as a clarifier.

Flocculation and sedimentation are generally highly effective at removing microorganisms. Guy et al.[8] reported that flocculation and sedimentation alone removed 99.9% of poliovirus and 99.7% of bacteriophage T4. Rao et al.[9] also reported that 98% of poliovirus, 95% of rotavirus, and 97% of hepatitis A were removed by these processes alone. However, other studies have not found anywhere near as good removal rates. For example, Payment et al.[10] reported only 77.8% reduction and Stetler et al.[11] reported a 50.9% reduction in enterovirus concentrations. One study even reported an increase in enteroviral levels during this stage of treatment.[12]

1.4.3 Filtration

After clarification, small flocs and other particulates are still suspended in the water. Most of the remaining particles are then removed during the process of filtration. Filtration involves passing the water through some porous medium such as sand. The particulates within the water are then trapped either within the pores of the filter medium or attached to the sand particles. Filters are made of a variety of materials such as sand, anthracite, or activated charcoal, either alone or in combination. Filters are classified according to the rate that water flows through them.

Perhaps the most common type of filter still in use is the rapid sand filter. In most circumstances, rapid sand filtration should only be used following coagulation and sedimentation. A rapid sand filter usually consists of 50 to 80 cm of coarse sand over a gravel, anthracite, or calcite base. Filters operate in the range of 5 to 24 $m^3/h/m^2$. Without prior treatment, pathogen removal can be quite poor. For example, rapid sand filtration alone only removed 1 to 50% of virus, but this was increased to over 99.7% after coagulation and settling.[13] As little as 40% of *Giardia* cysts are removed by rapid filtration alone.[14,15]

In some areas, slow sand filtration is still used. Slow sand filters consist of a layer of sand which is 60 to 120 cm deep over a graded gravel layer. The filtration rate is 0.1 to 1 $m^3/h/m^2$. When in use, a biologically active slime layer called the schmutzdecke develops on the open top 1 to 2 cm of the sand surface. This layer consists of filtered particulate matter and a complex living microbial community of bacteria, protozoa, algae, crustacea, and larvae. It is principally this biologically active layer which is responsible for removal of faecal microorganisms; physical filtration is less important. Slow sand filters can run for many weeks before clogging. The filter is then removed from service, drained, and cleaned by removing a few centimetres from the surface layer. The schmutzdecke layer must be re-established in a cleaned filter before it is returned to use; this can take several days. Properly run slow sand filters are highly effective in achieving a 4 to 5 log reduction of coliforms.[16] Over 99% of giardia cysts may be removed and over 99.99% of viruses.[17,18] Because of their low costs, slow sand filtration tends to be used for supplying smaller communities.

Other filter media include diatomaceous earth, the remains of siliceous shells of diatoms, and activated charcoal. After filtration, the water is usually disinfected.

1.4.4 Disinfection

Disinfection is the final barrier to the ingress of pathogenic microorganisms into the water supply. Several types of disinfectant and disinfectant systems are in use by the water industry (see Bitton for a more complete discussion[19]). Each disinfectant has its advantages and disadvantages making no one system ideal in all circumstances.

TABLE 1.1
Inactivation of Certain Pathogens by Chlorine at 5°C and pH = 6.0

Microorganism	Chlorine concentration mg/L	Inactivation time min
E. coli	0.1	0.4
Poliovirus 1	1.0	1.7
E. histolytica	5.0	18
G. lamblia	1.0	50
C. parvum	80.0	90

Source: Adapted from Hoff and Akin.[24]

The most common disinfectant is still chlorine. For larger treatment works, chlorine is nearly always added as the gas. When added to water, chlorine reacts to produce hypochlorous acid and hypochlorite. Sometimes chlorine is added indirectly by the addition of chlorinated lime (a solid) or sodium hypochlorite (a liquid). These are easier to use in smaller works but may lose their potency once opened.

Chlorine kills microorganisms by disrupting cell permeability and damaging nucleic acids and cellular enzymes. Unlike some other disinfectants, chlorine can remain at bactericidal levels in water distribution and maintain protection against regrowth of bacteria and possible contamination after the water leaves the treatment works.

Most pathogenic and indicator bacteria are killed fairly rapidly in the presence of chlorine of the same concentration as added to water supplies. However, certain viruses and protozoan cysts such as *Giardia* or *Cryptosporidium* survive substantially longer (Table 1.1). *Cryptosporidium*, in particular, is very resistant to standard concentrations of chlorine.

The main problem with chlorine is that, in certain waters, it can form trihalomethanes such as chloroform, 1,2-diclorethane, and carbon tetrachloride, and other disinfectant by-products. Trihalomethanes have been linked to both bladder and colon cancer.[20] It is for this reason that other disinfectants are being used more frequently, particularly in the U.S.

Chloramination is one method of avoiding the problem of chlorination by-products. In water hypochlorous acid reacts with ammonia to form chloramines. The exact proportion of each chloramine depends on the pH of the water with monochloramine predominating at pH greater than 6.0 and trichloramine predominating at pH less than 4.0 (Figure 1.4). Monochloramine is the preferred chloramine because the others can have an unpleasant taste. Chloramine does not have as rapid a bacteriocidal effect as does free chlorine, and some pathogens are quite resistant. However, chloramines are not inactivated as rapidly and thus can exert their antimicrobial activity for a longer period of time.

Chlorine dioxide is another disinfectant that is becoming more popular. It is formed by the reaction of chlorine gas and sodium chloride and is shown in the following equation

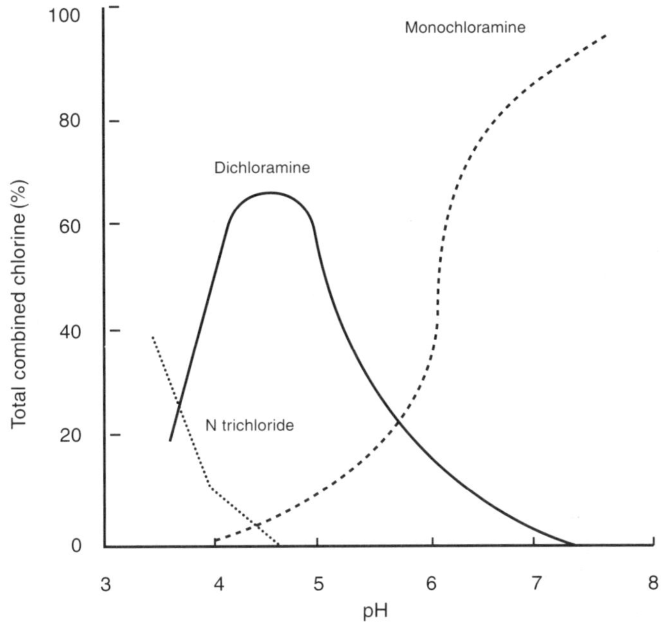

FIGURE 1.4 Distribution of types of chloramines with pH. (Courtesy of Wolfe, R.L., Ward, N.R., and Olson, B.H., Inorganic chloramines as drinking water disinfectants: a review, *J. Am. Water Works Assoc.*, copyright 1984. Material is reproduced with permission of The McGraw-Hill Companies.)

$$2\ NaClO_2 + 2\ Cl_2 \rightarrow ClO_2 + 2\ NaCl$$

Chlorine dioxide is a rapid and effective disinfectant. Chlorine may be added after the chlorine dioxide treatment to provide residual disinfectant in distribution.

The other main disinfectant in common use in commercial water treatment is ozone. Ozone is produced by passing dried air between two electrodes through which a high voltage alternating current is passed. Because of this demand for electricity, it is a more expensive form of disinfection. Also, no disinfection residual passes into the main supply so that bacterial regrowth is a potential problem unless another disinfectant is added. Compared to the other disinfectants discussed in this chapter, ozone has good activity against both *Cryptosporidium* and *Giardia*.[21,22]

Ultraviolet (UV) light has a long history in drinking water disinfection, although technical difficulties led it to be replaced by chlorine. However, recent technical improvements have led to its increased use in a variety of situations. It has great potential in potable water disinfection on both the small and large scale. Provided the system is installed and maintained adequately, UV disinfection does not produce any toxic by-products, taste, or odour problems. There is also no need to store dangerous chemicals. If the dose is adequate, *Cryptosporidium* and *Giardia* cysts are inactivated.

1.5 WATER DISTRIBUTION

After treatment, water is distributed to the eventual customer through a network of pumping stations, service reservoirs, water mains, and service pipes.

Service reservoirs are the slack in the system, enabling the system to cope with marked variations in demand throughout the day. Usually, the service reservoirs have a capacity equal to a little more than a single day's demand. They are constructed from concrete, brick, or steel and designed to be watertight in order to prevent the risk of contamination. In flat areas, these service reservoirs will often be built in water towers to give added hydraulic pressure within the distribution system. In some cases, secondary disinfection of the treated water is undertaken at the service reservoir to maintain residual amounts all the way to the point of delivery to the customer.

Trunk mains are the largest diameter main and are used for transporting large volumes of water over distances. They do not branch nor have connections to service pipes. The distribution mains are designed to distribute water from a supply reservoir to the individual customer's premises. They range in diameter from 50 to 500 mm and are made from a variety of materials including iron, asbestos cement, uPVC (unplasticized polyvinyl chloride), and MDPE (medium density polyethylene). Service pipes are the final stage in water distribution, taking the water from the distribution main into the customer's home.

The safety and quality of the distributed water depends on the integrity of the distribution system. Any break in that integrity could lead to contamination. For example, repair work on the distribution network may lead to ingress of contaminated water. There are national codes of practice to minimise such public health risks. Leakage from cracks and joints in older distribution pipes is known to occur. Generally, in a pressurised main, this does not present a risk because water will constantly leak outwards. However, if there is local loss of pressure owing to a nearby burst or sudden demand, for example, fire fighting purposes, the contaminated water can seep into the underground pipes. Loss of pressure can also result in back siphonage of contaminated water from incorrectly installed domestic appliances such as dishwashers and washing machines.

As will be seen in subsequent chapters, a variety of organisms can gain access to a distribution network where they can form biofilms on the pipe surfaces. Such biofilms can cause a variety of problems including deterioration in water taste, appearance, and quality, promote the survival of potentially pathogenic organisms, and otherwise interfere with the uses to which the water was intended.

The responsibility of the water companies ceases at the street boundary with the property which is supplied. The remainder of the connections and internal plumbing are owned by and are the responsibility of the customers, though water bylaws set out requirements for domestic plumbing. Deterioration of water quality can occur as water passes through pipework in a customer's property. This is a particular problem in large buildings such as hospitals with long pipe-runs.[23]

1.6 REFERENCES

1. **Anon.,** 1996, Water and sanitation, WHO Fact Sheet No. 112, World Health Organization, Geneva.
2. **Anon.,** 1998, *Microsoft Encarta Encycopaedia,* Microsoft.
3. Brockington, C. F., 1966, *In a Short History of Public Health*, Churchill J. and A. Ltd., London.
4. Hamann, C. L., Jr., McEwen, J. B., and Myers, A. G., 1990, Guide to selection of water treatment processes, in *Water Quality and Treatment: A Handbook of Community Water Supplies,* Pontius, F. W., Ed., McGraw-Hill, New York, 157.
5. Hunter, P. R. and Burge, S. H., 1988, Monitoring the bacteriological quality of potable waters in hospital, *J. Hosp. Infect.,* 12, 289.
6. Reinert, P. E. and Hroncich, J. A., 1990, Source water quality management, *Water Quality and Treatment: A Handbook of Community Water Supplies*, Pontius, F. W., Ed., McGraw-Hill, New York, 189.
7. Pedley, S. and Howard, G., 1997, The public health implications of microbiological contamination of groundwater, *Q. J. Eng. Geol.,* 30, 179.
8. Guy, M. D., McIver, J. D., and Lewis, M. J., 1977, The removal of virus by a pilot treatment plant, *Water Res.,* 11, 421.
9. Roa, V. C., Symons, J. M., Ling, A., Wang, P., Metcalf, T. G., Hoff, J. C., and Melnick, J. L., 1988, Removal of hepatitis A virus and rotavirus by drinking water treatment, *J. Am. Water Works Assoc.,* 80(2), 59.
10. Payment, P., Trudel, M., and Plante, R., 1985, Elimination of viruses and indicator bacteria at each step of treatment during preparation of drinking water at seven water treatment plants, *Appl. Environ. Microbiol.,* 49, 1418.
11. Stetler, R. E., Ward, R. L., and Waltrip, S. C., 1984, Enteric virus and indicator bacteria levels in a water treatment system modified to reduce trihalomethane production, *Appl. Environ. Microbiol.,* 47, 319.
12. Keswick, B. H., Gerba, C. P., DuPont, H. L., and Rose, J. B., 1984, Detection of enteric viruses in treated drinking water, *Appl. Environ. Microbiol.,* 47, 1290.
13. Robeck, G. G., Clarke, N. A., and Dostal, K. A., 1962, Effectiveness of water treatment processes for virus removal, *J. Am. Water Works Assoc.,* 54, 1275.
14. Logsdon, G. S., Symons, J. M., Hoye, R. L. J., and Arozarena, M. M., 1981, Alternative filtration methods for removal of cysts and cyst models, *J. Am. Water Works Assoc.* 73, 111.
15. Ongerth, J. E., 1990, Evaluation of treatment removing *Giardia* cysts, *J. Am. Water Works Assoc.* 82, 85.
16. Fox, K. R., Miltner, R. J., Logsdon, G. S., Dicks, D. L., and Drolet, L. F., 1984, Pilot-plant studies of slow-rate filtration, *J. Am. Water Works Assoc.,* 76, 62.
17. Bellamy, W. D., Silverman, G. P., Hendricks, D. W., and Logsdon, G. S., 1985, Removing *Giardia* cysts with slow sand filtration. *J. Am. Water Works Assoc.,* 77, 52.
18. Poynter, S. F. B. and Slade, J. S., 1977, The removal of viruses by slow sand filtration, *Prog. Water Technol.,* 9, 75.
19. Bitton, G., 1994, *Wastewater Microbiology,* Wiley-Liss, New York.
20. Hunter, P. R., 1997, *Waterborne Disease: Epidemiology and Ecology,* Wiley, Chichester, U.K.
21. Peeters, J. E., Mazas, E. A., Masschelein, W. J., Martinez de Maturana, I. V., and Debacker, E., 1989, Effect of disinfecting drinking water with ozone or chlorine dioxide on survival of *Cryptosporidium, Appl. Environ. Microbiol.,* 55, 1519.

22. Wickramanayake, G. B., Rubin, A. J., and Sproul, O. J., 1985, Effect of ozone and storage temperature on *Giardia* cysts, *J. Am. Water Works Assoc.,* 77, 74.
23. Hunter, P. R. and Burge, S. H., 1988, Monitoring the bacteriological quality of potable waters in hospital, *J. Hosp. Infect.,* 12, 289.
24. Hoff, J. C. and Akin, E. W., 1986, Microbial resistance to disinfectants: mechanisms and significance, *Environ. Health Perspect.,* 69, 7.
25. Davis, M. L. and Cornwell, D. A., 1991, *Introduction to Environmental Engineering,* McGraw-Hill, London.
26. Smethurst, G., 1988, *Basic Water Treatment,* 2nd ed., Thomas Telford, London.
27. Wolfe, R. L., Ward, N. R., and Olson, B. H., 1984, Inorganic chloramines as drinking water disinfectants: a review, *J. Am. Water Works Assoc.*, 76, 75.

2 Epidemiology

CONTENTS

2.1 INTRODUCTION

The assessment of risk in human populations relies heavily on the science of epidemiology. Epidemiology is that branch of medicine covering the study of the incidence and transmission of disease in populations. Thus the focus of study is groups of people rather than the individual. There are, as yet, no epidemiological studies that have specifically investigated the public health importance of biofilms in potable water systems. Nevertheless, epidemiological investigation is essential to the understanding of waterborne disease. In 1854, John Snow simultaneously invented the science of epidemiology and proved that drinking water was capable of spreading disease. Since then, epidemiological studies have been important in elucidating the role of water in many infectious diseases. In particular, epidemiological methods are central to the investigation of outbreaks of waterborne disease. Furthermore, as will be seen in a subsequent chapter, the results from epidemiological studies are essential in defining several of the input variables for various risk assessment models which are finding important roles in assessing the risks of waterborne disease.

By itself, epidemiology does not offer proof of the causation of disease. Instead, epidemiological investigations identify associations between disease and possible risk factors. At best, we are left with the demonstration of statistically significant associations between environmental factors and disease. Those factors found to be associated with disease may be causative, but other unknown confounders may be responsible for the association.

As with all sciences, epidemiology has developed its own jargon, the understanding of which is essential to the correct interpretation of the concepts being described. Some examples of the most important jargon are listed below:

Incidence The number of new cases occurring within a certain population during a specified time period. For example, we can state that the incidence of Salmonella in the U.K. is 23,400 cases per year.

Prevalence This is defined as the number of cases of a disease within a specified population at a specific point in time. For example, if one were to visit a village and count the number of children suffering from asthma, we could say that the prevalence of asthma in children under 10 was 10 out of 100 or 10%.

Absolute risk This is another way of referring to the incidence of disease. As a marker of risk, it does not give much information about the likely cause of disease.

Attributable risk The risk of disease owing to a specific environmental or other factor. We could say that of an incidence of 5000 cases per year, 20% were owing to contact with animals.

Relative risk Relative risk (*RR*) is the ratio between the incidence of disease in those members of the population exposed to a possible risk factor and those not exposed. Usually, 95% confidence intervals are also given to demonstrate the potential variability in the calculated ratio. (See Table 2.1 for the calculation of RR.)

TABLE 2.1
Calculation of Relative Risk

	Number of People Ill	Number of People Healthy	Total in Cohort
Exposure to risk factor	*A*	*B*	*A* + *B*
No exposure to risk factor	*C*	*D*	*C* + *D*

$$RR = \left(A/(A+B)\right)/\left(C/(C+D)\right) = \left(A*(C+D)\right)/\left(C/(A+B)\right)$$

Odds ratio The odds ratio (*OR*) is the ratio between the probability that someone with a disease has experience of the potential environmental factor and the probability that someone without the disease has experience of the same factor. Odds ratios are used when *RR* cannot be calculated because one does not know the real population incidence. They are primarily used in case control studies. (See Table 2.2 for the calculation of *OR*.)

Type I error Data from a random sample that show a statistically significant association or difference when no such association or difference exists in the population.

Type II error Data from a random sample that fails to show a statistically significant association or difference when such association or difference actually exists in the population.

TABLE 2.2
Calculation of Odds Ratio

	Cases	Controls
Exposure to risk factor	A	B
No exposure to risk factor	C	D

$$OR = (A/C)/(B/D) = AD/CB$$

2.2 EPIDEMIOLOGICAL METHODS AND WATERBORNE DISEASE

To many epidemiologists, modern epidemiology started with a waterborne outbreak of cholera linked to the Broad Street pump. What is more remarkable is that many of the skills that John Snow developed in 1849, such as case finding and accurate descriptive epidemiology, are still valid today. Although statistical techniques have come a long way since Snow's day, many would argue that the descriptive techniques he introduced are still the most valuable epidemiological investigations available.

There are a variety of types of epidemiological studies which have been used to investigate the relationship between water and disease. These epidemiological studies are generally divided into descriptive and analytical. Because any understanding of risk to health is based on epidemiology, it is important for the reader to have a basic understanding of these principles. To provide a context for these various studies, the following discussion of epidemiological methods form part of a broader discussion of an approach to investigating outbreaks.

The investigation of waterborne outbreaks, or any outbreak, should follow a logical process, which will be broadly similar in each case.

1. Preparation for the outbreak.
2. Detection of an outbreak.
3. Confirmation that an outbreak is occurring.
4. Description of the outbreak.
5. Generation of an hypothesis as to the cause of the outbreak.
6. Implementation of an initial control measure.
7. Testing the validity of the hypothesis.
8. Implementation of further control measures if necessary.
9. Learning the lessons for the future.

The best epidemiological investigations always follow this outline, although there is often much iteration and cycling back to earlier stages as new information becomes available or hypotheses are disproved. Often, different stages of the process are carried out at the same time. For example, the confirmation and description steps can happen at the same time. Hopefully, however, the preparation step will precede the others. Furthermore, the whole process can be thought of as cyclical rather than

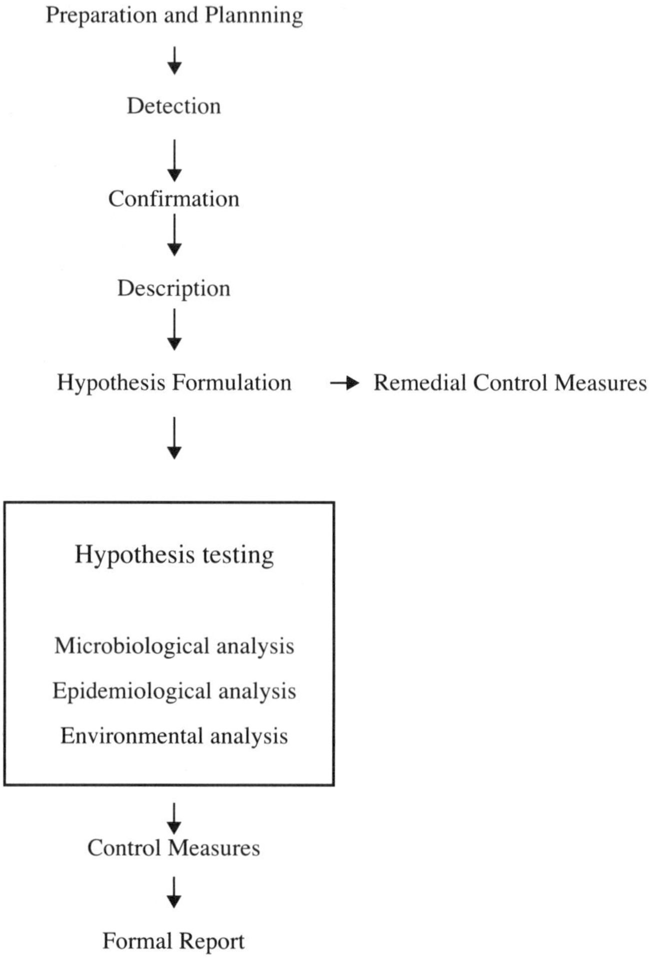

FIGURE 2.1 The stages in the investigation and control of a waterborne outbreak. (From *Water-Borne Disease: Epidemiology and Ecology,* Hunter, P. R., copyright © 1997. Reprinted by permission of Wiley-Liss, Inc., a subsidiary of John Wiley & Sons, Inc.)

linear. Ideally, the lessons learned in any outbreak should feed back into the planning for the next outbreak. Figure 2.1 gives an outline flow diagram of the stages in the investigation of an outbreak.

2.2.1 Preparation for the Outbreak

The investigation of waterborne outbreaks is a complex activity requiring various individuals from various organisations and professional backgrounds. For this group to function most effectively, everyone needs to know in advance what is expected of them. It is too late to decide who should be involved and what their respective roles in the investigation should be when the outbreak is already happening. All relevant organisations and authorities must have formal outbreak plans to which each have agreed previously. These outbreak plans must explicitly outline who is to manage the outbreak, who should be members of the outbreak team, and what is expected of them. In U.K. practice, the lead responsibility for the outbreak usually rests with the Consultant in Communicable Disease Control (CCDC) of the relevant health authority. Other members of the outbreak team will include a regional epidemiologist, medical microbiologist from the local hospital and/or Public Health Laboratory, environmental health officers from the local authority, one or more people from the water company, and a health or local authority press officer.

2.2.2 Detection of an Outbreak

Responsibility for surveillance of infectious disease rests with the CCDC. He or she will learn of an outbreak in one of various ways including analysis of routine notifications from general practitioners or microbiology laboratories. Alternately, the outbreak may be brought to his or her attention by the laboratory which has already detected the sudden increase. It must, however, be remembered that most reporting systems greatly underestimate the incidence of infectious diseases. Cases are lost throughout the chain of detection. Infected individuals may not develop symptoms or if they do, they may not present themselves to a doctor. If the patient gets to a doctor, the doctor may not make the correct diagnosis or send a specimen to the laboratory. The clinician may not remember to report the disease to the CCDC. The laboratory may not yet examine all faecal specimens for *Cryptosporidium* or if they do, the specimen may not be positive or the laboratory may not notify a positive result. Finally, the CCDC may not detect an increase in cases or realise the significance of a small increase in notifications. If the outbreak is spread over several authorities, the larger outbreak picture may not be detected. Each of these loss of information sources conspire together to ensure that probably most small outbreaks are not detected.

Rarely, outbreaks may be identified because a water company is aware of some treatment or distribution failure and advises the public health authorities who instigate prospective surveillance.

2.2.3 Outbreak Confirmation

Before acting on the diagnosis of an outbreak, it is important to find out whether the outbreak is real. Is the increase in cases an artefact as a result of increased testing or changes in reporting practices? Check that any microbiological diagnoses are correct. Is what is seen under the microscope really a *Cryptosporidium*? Have some family doctors suddenly realised they have to notify a particular disease? Such confirmation should not take long but could save a lot of embarrassment. Once the outbreak is confirmed, the outbreak team is called together.

2.2.4 Outbreak Description

The descriptive epidemiology of an outbreak is undoubtedly the most important stage in its investigation.[1] Descriptive epidemiology is essentially the description of the outbreak, giving the number of people affected, their characteristics, and when they became ill. Although descriptive epidemiology rarely proves disease associations, it is an essential starting point in the investigation of any outbreak or possible waterborne disease. Descriptive epidemiology will help generate hypotheses for further study.

More generally, descriptive studies set out to describe the pattern of disease within a specified community. These are primarily used for generating hypotheses of the cause of disease that can then be tested by subsequent analytical studies. They cannot be used to prove a disease association. However, any analytical study whose design is not based firmly on a preceding descriptive study is flawed.

The data used in descriptive studies usually come from routine surveillance data such as death reports, notifications of infectious disease, laboratory reports, or case-finding exercises. As such, data within descriptive studies are often incomplete because cases are not notified. The key items of information within any descriptive study are the temporal, geographic, and demographic distribution of disease, including such data as the date-of-onset, place of residence, travel history, age, sex, and food history of cases. Analysis is usually restricted to summarising and presenting this data in tabular and graphical form.

The first step in descriptive epidemiology is to agree to a case definition. This is essential in knowing whether individual cases of illnesses should be included in the outbreak. Case definitions may include a range of possible onset dates, clinical symptoms, geographical locations, and microbiological results. Case definitions can be very broad or very narrow to either include many possible cases or fewer cases. The broader the definition, the more cases will be identified, though many of these additional cases may not be related to the main outbreak. It is a matter of judgment how broad case definitions should be. A case definition does not have to be set in stone at this stage. It can and should change as new information becomes available.

Once a case definition has been agreed upon, case finding is the next step. For case definitions that include a microbiological diagnosis, the easiest way of identifying cases is to review microbiology laboratory results. A positive microbiological result will be very specific. However, relying on microbiological results will exclude those patients who have not had microbiology investigations taken. It is often necessary to encourage doctors to increase their sampling rate or report all episodes of particular clinical syndromes. A frequent alternative is to develop more than one case definition, one of which includes microbiology data and one that relies exclusively on clinical features. These can be called confirmed cases and presumptive cases.

A basic set of data needs to be collected on every individual who satisfies the case definitions. At a minimum, this will include name, address, age, sex, date of onset, the results of microbiological examination, and sufficient clinical information to prove that the individual satisfies the case definition. It is also usual to record place of work or schooling, a basic food or contact history, and any travel history. This type of data may be collected by a trawling questionnaire which asks a series of open questions covering activities during the period before the onset of illness. How far back the questioning should go depends on what is known about the incubation period of the particular disease under investigation.

FIGURE 2.2 Examples of the use of geographical mapping in the investigation of a waterborne outbreak. (Courtesy of Bridgman et al.[9] With permission from Cambridge University Press.)

The description of the outbreak then relies on the presentation of this early data such as the epidemic curve, geographical mapping, and age-sex distributions. The epidemic curve is probably the most useful descriptive technique in any outbreak investigation. It is simply a histogram showing the number of cases developing illness over time. Geographical mapping represents cases by marking points illustrating home addresses on appropriate maps such as street plans. When waterborne disease is suspected, the appropriate water supply zones may be superimposed on the map (Figure 2.2). Computer mapping packages are increasingly replacing the traditional pin-in-the-map technique.

Many epidemiological studies on water and health are of the ecological study. These studies are a variant of the descriptive study. In an ecological study, one attempts to draw conclusions on disease causation by correlating incidence or prevalence rates for several communities with possible factors such as the proportion drinking well water or proportion of unemployed at a community level. Ecological studies are flawed because they cannot link disease risk to individual exposure and cannot control for the effects of potentially confounding factors.[2] Furthermore, because ecological studies rely on average exposure levels, they may mask more complex relationships between exposure and disease. Despite their inherent weaknesses, they have been used extensively in studies of water and health and have frequently given misleading conclusions.[3]

Another form of descriptive study is the survey, although it is not used frequently in the investigation of outbreaks. Surveys seek to describe the characteristics of individuals in the population, including their personal attributes, experience of a particular disease, and exposure to putative causal agents. Surveys can be done in a variety of ways—by interviewing people in their homes or workplaces, telephone, and postal questionnaire. Usually, it is uneconomic and unnecessary, to interview all individuals in a population. In this case, only a proportion of the population is interviewed and a sample of the population is selected.

Samples can be selected in several ways. The easiest sampling technique is the random sample. Here, random number tables are used to choose individuals from a list of all such individuals in a population. Examples of such lists in the U.K. are electoral registers, telephone directories, and family doctors' lists of patients. Sometimes, it can be very difficult and time consuming to randomly select names from a register. In this case, a more convenient method is to select names at regular intervals from the list, say every tenth name on the electoral register or the first name on every page of the telephone directory. This is called a systematic sample.

Random sampling is not appropriate when the study is concerned with only a section of the population, such as children, women, or agricultural workers. In this case, one would choose a stratified sample. For example, a random search of doctors' lists could be restricted to certain age groups or particular locations. If social class is known to be significant, the random selection of individuals could be followed until the correct number of individuals from each social class has been recruited.

If one is interested in the prevalence of a disease in a region, it would be costly to select a truly random sample. A more economic sampling method in this situation is cluster sampling. Here, several villages may be randomly chosen for more detailed study. Sampling all the members of a random sample of households is another type of cluster sampling.

Whatever the sampling method employed, the major problem with surveys is one of bias, particularly selection bias. Selection bias may arise in one of three ways. The first is owing to the identification of the population. For example, using telephone directories as a source of names would exclude people who have no phone. This could have a significant effect on studies of any disease which tends to affect the poor and homeless. The second source of selection bias is deviation from the selection rules. If a field worker randomly is selecting individuals from a community, the selection process may be biased if he or she accepted volunteers rather than a truly random sample. The third way that a survey could be biased is if many originally selected individuals could not be traced or would not cooperate. For example, in a survey of diarrhoeal disease, those individuals who have been recently ill may be more likely to participate than those who have not been ill. Alternatively, if a high proportion of people with diarrhoea are admitted to the hospital, they may not be available for interview in the community.

2.2.5 Hypothesis Formulation

Once the initial outbreak data have been collected and described, the outbreak control team has to decide what possible factors may be responsible for the outbreak. In the context of waterborne disease, it has to be decided whether a waterborne factor is part of the initial hypothesis. In deciding what hypotheses to test, the outbreak control team will rely on the initial descriptive epidemiology, initial environmental investigations, knowledge of the epidemiology or microbiology of the causative agent, and their experience from previous outbreaks.

The epidemic curve will help distinguish between an outbreak owing to a single point source event, an outbreak owing to continued problems, and an outbreak owing to spread from person to person. Continued monitoring of the epidemic curve is also a good indicator as to whether control measures have been effective. The age and sex distribution is often helpful in suggesting possible hypotheses. Outbreaks of enteric disease spread by person-to-person transmission tends to affect younger children. However, high levels of immunity in the adult population can also skew the age distribution. Sex differences may reflect important occupational factors such as agricultural activities. The geographical distribution of cases, such as their restriction to certain water supply zones, is often one of the first indicators of a waterborne route of infection.

The early environmental investigations may also give important clues. Knowledge of water treatment failures in the days before an outbreak or early results of environmental microbiology will guide hypothesis generation.

Not to be underestimated, however, is the importance of experience in deciding what hypotheses to test. Nevertheless, always keep an open mind, as infectious disease epidemiology is usually able to surprise.

2.2.6 Remedial Control Measures

Once mains drinking water has been implicated as cause of an outbreak in the hypothesis generation phase of the investigation, the outbreak team has to decide whether any action should be taken at that stage to minimise any further hazard to the public. Essentially, the measures available are to disconnect the mains supply, use water from another source or reservoir, issue advice to the public to boil water for drinking, or do nothing.

It is this stage of outbreak investigation that is the most difficult. Rarely has the team proven the source of the outbreak. Furthermore, it is often not known whether any contamination was a short-lived or continuing problem. Decisions taken can have far reaching effects on individuals' health. Even issuing boil water notices can have their adverse health effects.[4] Needless to say, an innocent water company stands to lose money and reputation if it has to make major engineering changes and go public on a suggestion that is eventually disproved.

2.2.7 Hypothesis Testing (Analytical Epidemiology)

Once one or more hypotheses have been generated, the outbreak team then has to attempt to prove the hypotheses in a more rigorous fashion. There are usually three strands to this proof: microbiological, epidemiological, and environmental. Microbiological proof rests, at best, on identifying the causative agent in the water supply or demonstrating microbiological evidence of treatment failure such as coliforms. Microbiological examination of patients and contacts are also essential in many outbreaks to identify new cases.

Epidemiological studies that are used at this stage are analytical studies and are usually of the case-control study design or, less commonly, the retrospective cohort design.

A cohort study is a study of a group of individuals for whom exposure data is known. For outbreaks, this would almost invariably be the retrospective cohort design. Here, one would identify all individuals within the exposed group and interview as many as possible. In cohort studies, relative risk is used to compare the incidence of disease between those exposed and those not exposed to a potential causative agent. Because of the need to interview all members of the population, cohort studies are only suitable for investigating outbreaks in relatively small groups such as schools or small communities with a single water supply.

By contrast, a prospective cohort study identifies a group of individuals whose exposure to the factor under investigation is known. This group is followed over time to identify those who go on to develop disease. The prospective cohort study is the most powerful epidemiological method available as estimation of exposure is not biased by knowledge of who will get the disease. Unfortunately, if incidence is low or the incubation period prolonged, the study may need to be very large or continued over many years.

Case-control studies are retrospective studies of events that preceded the onset of disease in a group of individuals. They are the most common analytical epidemiological method in waterborne disease. Case control studies seek to test hypotheses by comparing the incidence of a preceding event in those with the disease (cases) with a group of individuals who do not appear to have the disease (controls). The key to success in case-control studies is the correct definition of cases and the selection of controls.

In order to know whether a particular individual should be included in a case-control study as a case, one needs a clear definition of what a case should be. This case definition may include clinical, epidemiological, and microbiological, or other laboratory features. Case definition will be discussed in more detail later on. In many waterborne outbreaks, it is likely that all individuals that satisfy the case definition can be included in the case-control study. However, provided sufficient numbers can be recruited, only a sample of cases may need to be interviewed. When only a sample of cases are included in the study, the sampling techniques available and their pitfalls are similar to those described previously under surveys. Frequently used cluster sampling methods include using only cases presenting to the hospital or cases living in selected villages.

The crucial point in case-control studies is that controls should be a random sample of the population from which the cases were selected. This is not as easy as it sounds. Traditionally, case control studies were done on matched pairs. Matched controls were chosen to be similar to cases on certain matching criteria such as age, sex, and location of residence. The weakness with matching controls is that any effect that the matching criteria may have on disease causation cannot then be tested.

If cases were taken from a subgroup of a population, then controls should also be selected from that same subgroup. For example, where cases of disease are presented to the hospital, one may use patients with an unrelated illness as controls. In the situation where cases are cluster sampled from certain villages, the controls should also be selected from those same villages. If all cases occur in children, then controls should also be children. Other ways of identifying controls include using family doctors' or health authorities' lists of patients, asking cases to nominate controls from their local areas, and selecting controls in a semirandom way from telephone directories. Each of these methods have their problems as discussed previously. In our experience, many cases are unwilling to name individual controls. If the number of cases is small, it is often necessary to interview more controls than cases to increase statistical sensitivity.

Because one cannot extrapolate from the results of a case-control study to the incidence of disease in the general population, odds ratios, rather than relative risks, are calculated.

Case control studies have the advantage of being relatively quick and inexpensive compared to some other designs of epidemiological study. They can also be used to examine several hypotheses. However, statistical association in a case control study does not necessarily imply causation. Care should be taken if many hypotheses are tested. If 20 variables are tested in a case control study and a p value of less than 0.05 is taken as indicating significance, it is highly probable that at least 1 variable will achieve statistical significance purely by chance. Furthermore, if the variables are not truly independent of one another, confounding variables may appear significant. These are some of the reasons why newer statistical techniques such as logistic regression analysis are becoming the favoured tools for analysing the results of analytical epidemiological studies.

Environmental investigations can also provide important proof. These investigations include a search for possible breakdowns in standard procedures, such as evidence of chlorination failure. This will include a review of all recent records relating to the quality and safety of the relevant supply along with a thorough inspection of the water source, treatment plant, and distribution network. In borderline cases, investigators may have to develop some particular risk assessment of the system by looking at its design and operating principles. Such environmental evidence both supports the other evidence and suggests control measures. Both U.S. and U.K. authorities have suggested protocols for determining the strength of evidence in favour of an outbreak being waterborne.[5,6]

Rarely do epidemiologists get the opportunity to do experiments into the causation of disease. In the field of waterborne disease, the situations where this may arise are monitoring the effects of health by putting a new water supply into an area previously without one, giving some households additional water treatment facilities, and temporarily stopping water supply or issuing a boil water notice during the investigation of a waterborne outbreak. When such opportunities arise, these offer unparalleled opportunities for studying the relationship between water supply and health or disease. There have been very few prospective experimental studies of recreational and potable waterborne disease reported in the literature to date.[7,8]

2.2.8 Control Measures

Once a waterborne hypothesis has been confirmed, one can have more confidence in the remedial measures already implemented. The water company can then institute more long-term control measures to stop a recurrence of the problem in the future. This may mean a change in procedure or new engineering works such as the introduction of new filtering systems.

2.2.9 Formal Report

In today's society, many different organisations and individuals will have an interest in the outcome of the investigation of an outbreak of waterborne disease. The water company and water regulators will need to know how the outbreak happened in order to reduce the risk of similar episodes happening in the future. The water regulator may be interested in knowing whether the failure was negligent if a prosecution was under consideration. Other water companies may want to put into effect the recommendations before the embarrassment of having an outbreak. Other public health teams may need to know the outcome of the investigations in case they also have a similar problem. Furthermore, local residents and the various enforcement authorities also need to know how the outbreak was handled and whether the control measures have been properly implemented. Not the least is to decide whether or not to seek legal redress for any harm or loss suffered from the outbreak. It is for all these people that a formal report needs to be written and published.

In order for the wider public health community to learn any general lessons, the appropriate national surveillance centre needs to know the outcome in order to record the nature and causes of the outbreak. Where more general lessons can be drawn, the essential aspects of the investigation should also be published in more widely read scientific journals so that authors can include the outbreak in their own reviews of the literature.

2.3 FURTHER READING

Farmer, R. and Miller, D., 1991, *Lecture Notes on Epidemiology and Public Health Medicine,* 3rd ed., Blackwell Scientific Publications, Oxford.

Hennekens, C. H. and Buring, J. E., 1987, *Epidemiology in Medicine,* Little, Brown, Boston.

Riegelman, R. K., 1981, *Studying a Study and Testing a Test, How to Read the Medical Literature*, Little, Brown, Boston.

2.4 REFERENCES

1. Palmer, S. R., 1989, Epidemiology in search of infectious diseases: methods in outbreak investigation, *J. Epidemiol. Community Health,* 43, 311.
2. Hennekens, C. H. and Buring, J. E., 1987, *Epidemiology in Medicine,* Little, Brown, Boston.
3. Hunter, P. R., 1997, *Water-Borne Disease: Epidemiology and Ecology,* Wiley, Chichester.
4. Mayon-White, R. T. and Frankenberg, R. A., 1989, Boil the water, *Lancet,* ii, 216.

5. Kramer, M. H., Herwaldt, B. L., Craun, G. F., Calderon, R. L., and Juranek, D. D., 1996, Surveillance for waterborne-disease outbreaks — United States, 1993–1994, *MMWR,* 45(SS-1), 1.
6. Tillett, H. E., de Louvois, J., and Wall, P. G., 1998, Surveillance of outbreaks of waterborne infectious disease: categorizing levels of evidence, *Epidemiol. Infect.,* 120, 37.
7. Fleisher, J. M., Jones, F., Kay, D., Stanwell-Smith, R., Wyer, M., and Morano, R., 1993, Water and non-water risk factors for gastroenteritis among bathers exposed to sewage-contaminated marine waters, *Int. J. Epidemiol.,* 22, 698.
8. Payment, P., Richardson, L., Siemiatycki, J., Dewar, R., Edwards, M., and Franco, E., 1991, A randomized trial to evaluate the risk of gastrointestinal disease due to consumption of drinking water meeting currently accepted microbiological standards, *Am. J. Public Health,* 81, 703.
9. Bridgman, S. A., Robertson, R. M. P., Syed, Q., Speed, N., Andrews, N., and Hunter, P. R., 1995, Outbreak of cryptosporidiosis associated with a disinfected groundwater supply, *Epidemiol. Infect.,* 115, 555.

3 Waterborne Diseases

CONTENTS

3.1 WATERBORNE DISEASE AND ITS EPIDEMIOLOGY

Since John Snow proved that drinking water could transmit cholera, many diseases have been shown to be spread by water. Estimates vary widely as to the actual morbidity and mortality owing to waterborne disease. The World Health Organization (WHO) estimates that every 8 seconds a child dies from a water-related disease and each year more that 5 million people die from illnesses linked to unsafe drinking water or inadequate sanitation.[1] Table 3.1 gives some estimates of the burden of disease associated with water-associated disease. While the figures in this table seem alarming, the situation is likely to deteriorate substantially as the world population continues to increase. The WHO also suggests that if sustainable safe drinking water and sanitation services were provided to all, each year there would be

200 million fewer diarrhoeal episodes.
2.1 million fewer deaths caused by diarrhoea.
76,000 fewer dracunculiasis cases.
150 million fewer schistosomiasis cases.
75 million fewer trachoma cases.

There are four ways by which water, or the lack of it, may be associated with disease.[1]

Waterborne diseases This is caused by the ingestion of water contaminated by human or animal faeces or urine containing pathogenic bacteria or viruses. It includes cholera, typhoid, amoebic and bacillary dysentery, and other diarrhoeal diseases.

Water-washed diseases This is caused by poor personal hygiene and skin or eye contact with contaminated water. It includes scabies, trachoma, and flea, lice, and tick-borne diseases.

Water-based diseases This is caused by parasites found in intermediate organisms living in water. It includes dracunculiasis, schistosomiasis, and other helminths.

Water-related diseases This is caused by insect vectors which breed in water. It includes dengue, filariasis, malaria, onchocerciasis, trypanosomiasis, and yellow fever.

TABLE 3.1
Estimates of Morbidity and Mortality of Water-Related Diseases

Disease	Morbidity (Episodes/Year, or as Stated)	Mortality (Deaths/Year)	Relationship of Disease to Water Supply and Sanitation
Diarrhoeal diseases	1,000,000,000	3,300,000	Strongly related to unsanitary excreta disposal, poor personal and domestic hygiene, unsafe drinking water
Infection with intestinal helminths	1,500,000,000[1]	100,000	Strongly related to unsanitary excreta disposal, poor personal and domestic hygiene
Schistosomiasis	200,000,000[1]	200,000	Strongly related to unsanitary excreta disposal and absence of nearby sources of safe water
Dracunculiasis	100,000[2]	—	Strongly related to unsafe drinking water
Trachoma	150,000,000[3]	—	Strongly related to lack of face washing, often owing to absence of nearby sources of safe water
Malaria	400,000,000	1,500,000	Related to poor water management, water storage, operation of water points, and drainage
Dengue fever	1,750,000	20,000	Related to poor solid wastes management, water storage, operation of water points, and drainage
Poliomyelitis	114,000	—	Related to unsanitary excreta disposal, poor personal and domestic hygiene, and unsafe drinking water
Trypanosomiasis	275,000	130,000	Related to the absence of nearby sources of safe water
Bancroftian filariasis	72,800,000[1]	—	Related to poor water management, water storage, operation of water points, and drainage
Onchocerciasis	17,700,000[1,4]	40,000[5]	Related to poor water management in large-scale projects

[1] People currently infected.
[2] Excluding Sudan.
[3] Case of the active disease; approximately 5,900,000 cases of blindness or severe complications of Trachoma occur annually.
[4] Includes an estimated 270,000 blind.
[5] Mortality caused by blindness.
Source: WHO data.

In this book, we are primarily concerned with waterborne disease transmitted by drinking water. Table 3.2 lists those infectious diseases that are associated with drinking water and describes some of the symptoms that they cause. For a more detailed description of these diseases see *Waterborne Disease* by Hunter.[2]

TABLE 3.2
Microbial and Parasitic Disease Linked to Drinking Water Consumption

Disease	Organism	Usual Incubation Period	Symptoms
Dracunculiasis	*Dracunculus medinensis*	8–10 months	Painful ulcers on lower limbs and feet
Giardiasis	*Giardia duodenalis* (Prev. *G. lamblia*)	7–14 days	Diarrhoea, abdominal cramps, bloating and flatulence; for more prolonged disease, weight loss and failure to thrive
Cryptosporidiosis	*Cryptosporidium parvum*	7–10 days	Diarrhoea often lasting for several weeks
Cyclosporiasis	*Cyclospora cayetanensis*	1–7 days	Diarrhoea, abdominal pain, nausea, vomiting, and anorexia; weight loss in prolonged cases
Amoebiasis	*Entamoeba histolytica*	2–4 weeks	Varies from mild diarrhoea to a fulminating dysenteric illness
Toxoplasmosis	*Toxoplasma gondii*	Varies	A glandular fever-like syndrome; in pregnant women, can cause damage to foetus including abortion, hydrocephalus, cerebral calcification, and eye damage
Cholera	*Vibrio cholerae*	1–3 days	Painless watery diarrhoea which in severe cases can lead to dehydration, shock, and death
Salmonellosis	*Salmonella* spp.	12–48 hours	Diarrhoea, colicky abdominal pain, and fever; may progress to more severe systemic disease in a small proportion of cases
Typhoid	*Salmonella typhi*	10–14 days	Fever, malaise, and abdominal pain; as disease progresses may develop delerium; untreated death rate is up to 15%
Shigellosis	*Shigella* spp.	1–3 days	Ranges from mild self-limiting diarrhoea to more severe diarrhoea with painful straining to empty bowels, blood loss leading to collapse and death
Campylobacteriosis	*Campylobacter* spp.	2–4 days	Diarrhoea which may be bloody and cramping abdominal pain
Enterotoxigenic *E. coli*	*E. coli*	12–72 hours	Watery diarrhoea
Enterohaemorrhagic *E. coli*	*E. coli* 0157 and others	3–4 days	Bloody diarrhoea which can be fatal and progress to haemolytic uraemic syndrome in children
Yersiniosis	*Yersinia* spp.	3–7 days	Fever, diarrhoea, and abdominal pain

continued

TABLE 3.2 (continued)
Microbial and Parasitic Disease Linked to Drinking Water Consumption

Disease	Organism	Usual Incubation Period	Symptoms
Aeromoniasis	*Aeromonas* spp.	Uncertain	Diarrhoea
Tularaemia	*Francisella tularensis*	2–5 days	Either typhoid-like or mucocutaneous with suppurative skin lesions
Gastritis/ulceration	*Helicobacter pylori*	Varied	Prolonged gastritis, peptic ulcer, and gastric cancer
Viral hepatitis	Hepatitis A virus Hepatitis E virus	2–4 weeks 6–8 weeks	Mild flu-like symtoms to severe fulminating hepatitis and death; death is especially common with Hepatitis E in pregnant women
Viral gastroenteritis	Small round viruses Rotavirus	12 to 72 hours	Vomiting and/or diarrhoea
Poliomyelitis	Poliovirus types 1 to 3	9–12 days	Mostly assymptomatic though can progress to asceptic meningitis, encephalitis, and paralysis
Enteroviral illness	Coxsackieviruses A Echoviruses Enteroviruses	2–5 days	Various including diarrhoea, pneumonitis, and myalgia

For the rest of this chapter, we shall consider the question of how common is disease caused by potable water. We shall restrict our discussion to those studies done in the Western world. Essentially, evidence on the epidemiology of waterborne disease in the West comes from two sources, reports of waterborne disease and prospective studies of sporadic disease.

3.2 REPORTED OUTBREAKS OF WATERBORNE DISEASE

Very few countries have satisfactory surveillance systems for waterborne disease. Only the U.S. and U.K. have established surveillance systems with regular publication of details on waterborne outbreaks. In the U.S., the Center for Disease Control (CDC) has been collating and reporting on waterborne disease since 1971. In the U.K., the Public Health Laboratory Service Communicable Disease Surveillance Centre (CDSC) has been publishing biannual reports since 1994. Although data from both countries has been collected since before these schemes were implemented, data collection was less systematic. The two systems differ from each other, each having its own strengths and weakness. The U.S. system is probably more comprehensive including chemical incidents and many outbreaks of unknown aetiology. American citizens appear more likely to contact their health departments should they suffer from a gastrointestinal illness. By contrast, in the U.K. system, which is based on reporting of laboratory isolations, cases are usually only identified after

TABLE 3.3
Waterborne Outbreaks Associated with Public Water Supplies in England and Wales 1911–1998, Number of Outbreaks in 10-Year Periods

10-Year Period	Number of Outbreaks		Number of Cases and Deaths	Disease: Number of Outbreaks, Cases, and Deaths
	Site of Contamination	Source Distribution		
1911–1920	3	5	3630+	Typhoid[b]: 6, 359+, 28+
			28+	Paratyphoid: 1, 71, ?
				Dysentery: 1, 1700, ?
1921–1930	5	3	2029+	Typhoid: 4, 459, 50
			65+	Paratyphoid: 1, 31, ?
				Dysentery: 1, 1100, 12
				Gastroenteritis: 2, 439, ?3
1931–1940	6	0	7912+	Typhoid: 3, 686, 77+
			78+	Gastroenteritis: 3, 7200+, 1
1941–1950	1	2	610	Typhoid: 1, 22, 0
			0	Dysentery: 2, 588, 0
1951–1960	Nonidentified			
1961–1970	1	0	90	Paratyphoid: 1, 90, 0
			0	
1971–1980	1	2	3222	Gastroenteritis: 2, 3114, 0
			0	Giardiasis: 1, 60, 0
1981–1990	13	3	1925	*Campylobacter*: 3, 629, 0
			0	*Cryptosporidiosis*: 10, 857+, 0
				Gastroenteritis: 3, 310, 0
1991–1998	22	0	2550+	*Cryptosporidiosis*: 22, 2550+, 0
			0	

[a] Case numbers are often estimates, + indicates a minimum estimate.

[b] One outbreak of typhoid was also associated with 1500 cases of gastroenteritis.

Source: Adapted from Galbraith[11] with additional data from Stanwell-Smith,[12] Furtado et al.,[13] and various Communicable Disease Reports.[a]

they have been attended by doctors and samples have been taken for analysis. On the other hand, once a patient attends a medical practitioner, a sample is more likely to be sent for examination and once positive, reported to the surveillance systems. In any event, waterborne outbreaks are probably significantly underreported in both countries.

Reports of waterborne outbreaks in England and Wales are presented in Tables 3.3 and 3.4. Table 3.3 gives outbreaks related to mains water and Table 3.4 to private supplies. In the U.K. data, it is clear that there has been a significant change in the identified causes of waterborne disease during this century. Up until 1970, waterborne disease was dominated by typhoid or paratyphoid. Dysentery was

TABLE 3.4
Waterborne Disease from Private Supplies in England and Wales from 1941–1998 and Various Communicable Disease Reports

10-Year Period	Number of Outbreaks	Number of Cases and Deaths	Disease: Number of Outbreaks, Cases, and Deaths
1941–1950	4	47+	Typhoid: 2, 9+, 5
		5	Paratyphoid: 1, 21, 0
			Amoebiasis: 1, 17, 0
1951–1960	None detected		
1961–1970	None detected		
1971–1980	2	166	Paratyphoid: 1, 6, 0
		0	Gastroenteritis: 1, 160, 0
1981–1990	5	962	Gastroenteritis: 1, 138, 0
		0	*Campylobacter*: 3, 520, 0
			Streptobacillary fever: 1, 304, 0
1991–1998	13	382	Gastroenteritis: 2, 81, 0
		0	*Campylobacter*: 6, 147, 0
			Giardia: 1, 31, 0
			Cryptosporidiosis: 2, 66, 0
			E. coli: 1, 14, 0
			Mixed *Campylobacter* and *Cryptospridiosis*: 1, 43, 0

Source: Adapted from Galbraith et al.,[11] Stanwell-Smith,[12] and Furtado et al.[13].

the only other infection of note during this period. Since the 1970s, reports of waterborne outbreaks have increased substantially, although this increase has been essentially owing to *Cryptosporidium* in public water supplies and *Campylobacter* in private supplies. Both pathogens are relatively newly described with *Campylobacter* being described in the late 1970s and *Cryptosporidium* in the early 1980s. The decline in the reporting of typhoid, a disease with a high mortality without adequate treatment, reflects improved water treatment and the disappearance of the pathogen from the general population. The increase in the new pathogens reflects our ability to diagnose them. The relative paucity of *Campylobacter* in public compared to private water supplies reflects this pathogen's sensitivity to disinfection. In the U.K., virtually all public supplies have some form of disinfection, usually with chlorine. *Cryptosporidium* is relatively resistant to disinfection, hence its prevalence in public supplies. Although there are much fewer outbreaks reported from private supplies, a very small proportion of the population of England and Wales have a private supply. Also, given their smaller size, private supply outbreaks are probably more likely to be missed. Consequently, the relative risk of being involved in a waterborne outbreak is probably much higher in people drinking private water.

The U.S. data are shown in Table 3.5. In many ways, the U.S. data show the same trend over the century, as does the U.K. data. Over the years, typhoid declines in importance from its once preeminent position. There are, however, some major

TABLE 3.5
Aetiology of Waterborne Outbreaks in the U.S., 1920–1996

Time Period	Disease	Number of Outbreaks	Number of Cases
1920–1940	Typhoid	372	13,767
	Gastroenteritis	144	176,725
	Shigellosis	10	3,308
	Amoebiasis	2	1,416
	Hepatitis	1	28
	Chemical poisoning	1	92
	Subtotal	530	195,336
1941–1960	Gastroenteritis	265	54,439
	Typhoid	94	1,945
	Shigellosis	25	8,951
	Hepatitis	23	930
	Salmonellosis	4	31
	Chemical poisoning	4	44
	Paratyphoid	3	19
	Amoebiasis	2	36
	Tularaemia	2	6
	Leptospirosis	1	9
	Poliomyelitis	1	16
	Subtotal	424	66,426
1961–1970	Gastroenteritis	39	26,546
	Hepatitis	30	903
	Shigellosis	19	1,666
	Typhoid	14	104
	Salmonellosis	9	16,706
	Chemical poisoning	9	46
	Toxigenic *E. coli*	4	188
	Giardiasis	3	176
	Amoebiasis	3	39
	Subtotal	130	46,374
1971–1990	Gastroenteritis	293	67,367
	Giardiasis	110	26,531
	Chemical poisoning	55	3,877
	Shigellosis	42	8,947
	Viral gastroenteritis	27	12,699
	Hepatitis A	25	762
	Salmonellosis	12	2,370
	Campylobacteriosis	12	5,233
	Typhoid	5	282
	Cryptosporidiosis	2	13,117

continued

TABLE 3.5 (continued)
Aetiology of Waterborne Outbreaks in the U.S., 1920–1996

Time Period	Disease	Number of Outbreaks	Number of Cases
	Yersiniosis	2	103
	Toxigenic *E. coli*	1	1,000
	E. coli O157	1	243
	Chronic gastroenteritis	1	72
	Cyclosporiasis	1	21
	Cholera	1	17
	Amoebiasis	1	4
	Subtotal	591	142,645
1991–1996	Gastroenteritis	38	14,228
	Chemical poisoning	20	486
	Giardiasis	11	1,967
	Cryptosporidiosis	8	406,822
	Shigellosis	7	592
	Campylobacteriosis	3	223
	Hepatitis A	2	56
	E. coli 0157	2	35
	Salmonellosis	1	625
	Viral gastroenteritis	1	146
	Plesiomonas	1	60
	Vibrio cholera non 01	1	11
	Subtotal	95	425,251

Source: Adapted from Craun[14] with additional data from Moore et al.,[15] Kramer et al.,[16] and Levy.[17]

differences between the U.S. and U.K. data. The first thing to note is the much larger number of outbreaks in the U.S., even accounting for the larger population. In part this is owing to the American detection of outbreaks where no pathogen was isolated, but it also probably relates to the different nature of the water supply industry in the U.S. There are many more isolated small communities with their own water supplies in the U.S. than in the U.K. Many of these smaller supplies get little or no treatment. The other important difference is the large number of *Giardia* cases in the U.S. It is not totally clear why the U.S. sees so many *Giardia* cases compared to the U.K., although we suspect part of the explanation relates to the reported prevalence of *Giardia* in wild animals, who then contaminate small water supplies.

One advantage of the U.S. presentation is a more detailed analysis of the treatment failures which gave rise to the outbreaks. It can be seen from Table 3.6 that the vast majority of non-community waterborne outbreaks were owing to untreated or inadequately chlorinated groundwater. For the larger community outbreaks, the causes of failure were more diverse, including inadequate disinfection of surface water, distribution deficiencies, groundwater problems, and filtration deficiencies.

TABLE 3.6
Causes of Waterborne Outbreaks, U.S., 1981–1990

Cause of Outbreak	Community	Noncommunity	Other
Untreated groundwater	15 (12.1)	43 (44.3)	19 (27.1)
Inadequate disinfection of groundwater	17 (13.7)	32 (33.0)	—
Ingestion of contaminated water while swimming	—	—	41 (58.6)
Inadequate disinfection of surface water	35 (28.2)	9 (9.3)	—
Distribution deficiencies	30 (24.2)	3 (3.1)	3 (4.3)
Filtration deficiencies	16 (12.9)	1 (1.0)	—
Unknown	7 (5.6)	3 (3.1)	1 (1.4)
Untreated surface water	2 (1.6)	4 (4.1)	3 (4.3)
Miscellaneous	2 (1.6)	2 (2.1)	3 (4.3)
Total	124	97	70

Source: Reprinted from Craun.[14]

Although outbreaks are useful in identifying the microbiological and engineering causes of waterborne disease, they cannot give an accurate estimate of the burden of waterborne disease in a community. Many outbreaks will go undetected. To gain an insight into overall disease burden, we have to turn to other epidemiological methods. In the next section we will consider some of the epidemiological studies which have investigated the relationship between disease and water consumption.

3.3 EPIDEMIOLOGICAL STUDIES OF WATERBORNE DISEASE

There is very comprehensive literature on the effects of drinking water quality on health. Unfortunately, for this review, the vast majority concerns studies undertaken in developing countries.[3] By contrast, there have been very few studies in the Western world. There are essentially two approaches. The first approach is to conduct case-control studies on diagnosed sporadic infection. Clearly, this approach reveals information on just a small proportion of infections, those owing to the specific disease under investigation. Consequently, we have to rely on prospective studies of illness rates, either in cohort studies or experimental studies.

Of the prospective cohort studies of drinking water and ill health, in our view, the work of a group of French researchers is preeminent. Their first work was a prospective longitudinal study over 18 months of 52 French Alpine villages.[4,5] They collected data on water quality for each village and recorded the number of patients who were diagnosed with gastroenteritis each week by their physician. All villages were supplied with untreated surface water. Those villages whose water did not meet the European standard had a significant excess of cases of gastroenteritis (*RR* equal to 1.36, 95% CI 1.24 to 1.49). The most predictive marker of illness was faecal streptococci, although faecal coliforms were also independently associated. By contrast, total coliforms and aerobic plate counts were not independently associated with risk.

In a subsequent study, the group looked at the effect on health of chlorination of substandard water.[6] In much the same way as they had done in their previous study, they recorded the rate of diarrhoea in some 2033 school children living in 24 French villages. In these 24 villages, 13 did not give their water any treatment as it already met statutory standards. The other villages chlorinated their water but did no other treatment. Those children living in the villages with initially substandard water were 1.4 times more likely to suffer from a diarrhoeal illness (95% CI 1.30 to 1.50). Interestingly, this excess risk was associated with the occurrence of small epidemics. Thus faecally polluted water continued to pose an excess risk even after chlorination.

An interesting approach to the issue of trying to quantify risk to health from water was developed by two groups who looked at the temporal correlation between physician reported gastrointestinal illness rates and drinking water turbidity.[7,8] The first group looked at historical data correlated over time from January 1992 to April 1993 from Milwaukee County, WI (better known as the place where the world's biggest documented waterborne outbreak occurred).[7] The authors noted that a rise in turbidity of 0.5 NTU was associated with a 2.35 (95% CI 1.34 to 4.12) increase risk of gastroenteritis in children and 1.17 (0.91 to 1.52) in adults. The second group looked at the relationship between emergency visits and admissions to the Children's Hospital in Philadelphia for gastrointestinal illness. The authors reported that an interquartile increase in drinking water turbidity was associated with a 9.9% (2.9 to 17.3%) increase in visits in children, aged 3 years and over, 4 days later, a 5.9% (0.2 to 12%) increase 10 days later in children 2 years and younger. Hospital admissions followed the same trend.

In the first study of its kind, Payment and co-workers[8] randomly allocated point-of-use reverse-osmosis filters to one half of a study population in Montreal. Volunteers kept a health diary and the researchers were able to compare self-reported episodes of gastroenteritis in the two groups. Throughout the study period, rates of gastroenteritis were significantly higher in the group drinking unfiltered tap water, although laboratory investigations were unable to identify any pathogen responsible for this excess. The authors estimated that about 30% of all cases of gastrointestinal infections were attributable to the drinking water.

In a subsequent experimental study, Payment and colleagues compared randomly allocated volunteers to one of four study arms: tap water, tap water from a continuously running tap, bottled plant water, and purified bottled plant water.[9] The results of this study was much less convincing than the previous one, although the authors still felt able to suggest that 14 to 40% of gastrointestinal illnesses were related to drinking water. However, a major problem with Payment's studies were that all his volunteers knew which arm of the study they were in and thus the outcome of both his studies could have been affected by reporting bias on the part of his volunteers. At the time of publication, variations of Payment's studies were being repeated with the study design, ensuring that volunteers would be unaware of whether they were drinking filtered or unfiltered water. These studies should be reported during the year 2000 or 2001.

All the epidemiological studies reported previously have reported evidence for the association between drinking water and gastrointestinal illness. It would appear that even the drinking water meeting current microbiological standards might be associated with illness. It is still difficult to extrapolate from these studies some national estimate of disease burden related to mains drinking water consumption. The most extreme estimate from Payments work is 30% of all cases of gastrointestinal infections.

Another approach to estimating disease burden is to determine the prevalence of various infectious diseases and then estimate the proportion owing to water. Morris and Levin[10] has done this for the U.S. and produced estimates of between 520,000 and 690,000 (midpoint 560,000) cases of moderate to severe infection owing to water consumption and 400,000 to 27,000,000 (midpoint 7,100,000) cases of mild to moderate infection. Despite the amount of analysis that went into this work, many of the important variables had to be estimated. Consequently, the final estimates at best represent only an educated guess.

We cannot rely on currently available information to give reliable estimates of the total disease burden owing to mains drinking water. We know from outbreaks that there is a positive association between drinking water and some disease. More detailed estimates of sporadic disease will require further epidemiological work.

3.4 REFERENCES

1. **Anon.,** 1996, Water and sanitation, WHO Fact Sheet No. 112, World Health Organization, Geneva.
2. Hunter, P. R., 1997, *Waterborne Disease: Epidemiology and Ecology,* Wiley, Chichester.
3. Esrey, S. A., 1996, Water, waste, and well-being: a multicountry study, *Am. J. Epidemiol.,* 143(6), 608.
4. Ferley, J. P., Zmirou, D., Collin, J. F., and Charrel, M., 1986, Etude longitudinale des risques liés à la consommation deaux non conformes aux normes bactériologiques, *Rev. Epidemiol. Sante Publ.,* 34, 89.
5. Zmirou, D., Ferley, J. P., Collin, J. F., Charrel, M., and Berlin, J., 1987, A follow-up study of gastro-intestinal diseases related to bacteriologically substandard drinking water, *Am. J. Public Health,* 77, 582.
6. Zmirou, D., Rey, S., Courtois, X., Ferley, J. P., Blatier, J. F., Chevallier, P., Boudot, J., Potelon, J. L., and Mounir, R., 1995, Residual microbiological risk after simple chlorine treatment of drinking ground water in small community systems, *Eur. J. Public Health,* 5, 75.
7. Morris, R. D., Naumova, E. N., Levin, R., and Munasinghe, R. L., 1996, Temporal variation in drinking water turbidity and diagnosed gastroenteritis in Milwaukee, *Am. J. Publich Health,* 86, 237.
8. Payment, P., Richardson, L., Siemiatycki, J., Dewar, R., Edwards, M., and Franco, E., 1991, A randomized trial to evaluate the risk of gastrointestinal disease due to consumption of drinking water meeting currently accepted microbiological standards, *Am. J. Public Health,* 81, 703.
9. Payment, P., Siemiatycki, J., Richardson, L., Renaud, G., Franco, E., and Prévost, M., 1997, A prospective epidemiological study of gastrointestinal health effects due to the consumption of drinking water, *Int. J. Environ. Health Res.,* 7, 5.
10. Morris, R. D. and Levin, R., 1995, Estimating the incidence of waterborne infectious disease related to drinking water in the United States, in *Assessing and Managing Health Risks from Drinking Water Contamination,* Reichard, E. G. and Zapponi, G. A., Eds., International Association of Hydrological Sciences, Wallingford, U.K., 75.

11. Galbraith, N. S., 1994, A historical review of microbial disease spread by water in England and Wales, in *Water and Public Health,* Golding A. M. B., Noah, N., and Stanwell-Smith, R., Eds., Smith-Gordon & Co., London, 15.
12. Stanwell-Smith, R., 1994, Water and public health in the United Kingdom. Recent trends in the epidemiology of water-borne disease, in *Water and Public Health,* Golding A. M. B., Noah, N., and Stanwell-Smith, R., Eds., Smith-Gordon & Co., London, 39.
13. Furtado, C., Stuart, J. M., Adak, G. K., Evans, H. S., Knerer, G., and Casemore, D. P., 1996, Waterborne outbreaks of gastroenteritis in England and Wales: a four year review, in *PHLS 21st Annual Scientific Conference,* University of Warwick, PHLS, London, Poster No. 22.
14. Craun, G. F., 1992, Waterborne disease outbreaks in the United States of America: causes and prevention, *World Health Stat. Q.*, 45, 192.
15. Moore, A. C., Herwaldt, B. L., Craun, G. F., Calderon, R. L., Highsmith, A. K., and Juranek, D. D., 1993, Surveillance for waterborne disease outbreaks — United States, 1991–1992, *MMWR,* 42 (SS-5), 1.
16. Kramer, M. H., Herwaldt, B. L., Craun, G. F., Calderon, R.L., and Juranek, D. D., 1996, Surveillance for waterborne-disease outbreaks — United States, 1993–1994, *MMWR,* 45 (SS-1), 1.
17. Levy, D. A., Bens, M. S., Craun, G. F., Calderon, R. L., and Herwaldt, B. L., 1996, Surveillance for waterborne-disease outbreaks — United States, 1995–1996, *MMWR,* 47 (SS-5), 1.

4 Risk Assessment

CONTENTS

4.1 INTRODUCTION

Risk assessment has become increasingly important to public health in recent years. Indeed, the requirement to do formal assessments of the risk to water supply from cryptosporidiosis has recently been enshrined in British Law.[1] When one considers the risk to public health from biofilm formation, there are very few sources of direct evidence. Consequently, one has to use one or more of the various models of risk assessment. The National Academy of Sciences defined risk assessment as the use of the factual base to define the health effects of exposure of individuals or populations to hazardous materials and situations.[2] This definition does not, however, give much insight into the nature of risk and the processes involved in risk assessment. A further problem is that much of the information to do adequate risk assessments does not exist.

This chapter discusses risk assessment by first defining risk and looking at the role of risk assessment and then by describing some approaches to risk assessment that are applicable to the water industry.

4.2 DEFINITION OF RISK

Risk means very different things to different people. In the chapter on epidemiology, we have already come across several definitions of risk. More generally, risk was defined as the number of new cases of disease in a population of a known size over a given period of time (i.e., the incidence of disease). We also defined relative risk (the ratio of incidence in an exposed and nonexposed population) and attributable risk (the incidence owing to exposure to a specific factor).

More generally, risk can be defined as the possibility of loss, harm, or injury. This definition contains within it two concepts, harm and probability. This brings us to our first risk assessment model. Risk can be thought of as a two-dimensional matrix, which includes probability and severity of harm. This is shown graphically

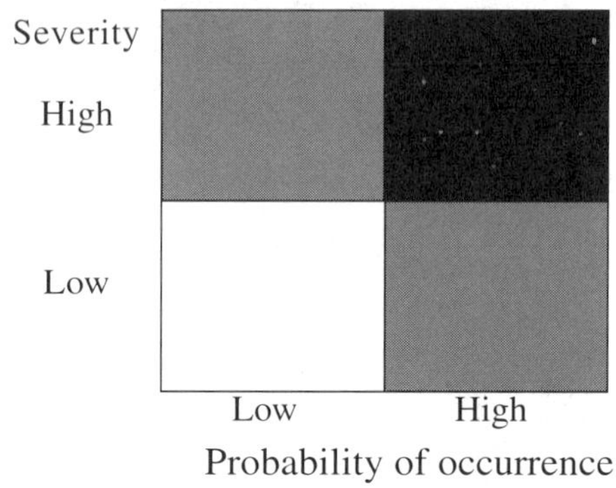

FIGURE 4.1 Probability/severity matrix.

in Figure 4.1. Clearly, any potentially adverse event can be placed on this matrix given its known or estimated probability of occurrence and severity. Once placed on the matrix, it is easier to identify those risks that need to be addressed most urgently (high probability, high severity). Low probability–low severity risks have low priority and the other classifications have intermediate priority.

The advantage of this particular model is that it is conceptually simple and can be used in developing inputs to a variety of other risk assessment models. It is also very good for the rapid modelling of a large number of potential hazards. On the other hand, the exact probabilities are often unknown and thus the placing of individual hazards into the matrix still has a large amount of subjectivity. Although rarely explicitly used in this form, this model is central to all risk assessments. In the authors' view, the value of risk assessments would be enhanced if this model were used explicitly more often.

4.2.1 Role of Risk Assessment Models

Now that we have been introduced to the concept of risk and one risk assessment model, we should consider the role of risk assessment, particularly in the water industry. There have been many different reasons put forward for undertaking risk assessment. Nevertheless, the main reason should be to improve the safety of our water supplies. To be most effective, risk assessment must be part of a process that leads to risk communication and, in turn, risk reduction.

Some of the many potential uses of risk assessment in the water industry include

To predict the burden of waterborne disease in the community under outbreak and nonoutbreak conditions. This is helpful in determining the impact of improvements in water supply on health and acts as a driver toward technological improvement.

To help set microbiological standards for drinking water supplies that will give acceptable levels of illness within the populations drinking that water.

To identify the most cost-effective option to reduce microbiological health risks to drinking-water consumers.

To help determine the optimum treatment of water to balance microbial risks against chemical risks from disinfection by-products.

To provide a conceptual framework to help individuals and organisations understand the nature and risk to and from their water and how those risks can be minimised.

4.2.2 Epidemiological Risk Assessment

We have already mentioned the epidemiological definitions of risk, relative risk, and attributable risk. Defining these risks requires one or more epidemiological studies. Once the risks have been assessed by epidemiological investigations, the information produced can be thought of as the Holy Grail of risk assessment. Unfortunately, as with the legendary Holy Grail, the goal is rarely achieved. Epidemiological data is often costly to collect, requiring descriptive studies, surveys, or cohort studies. Furthermore, all epidemiological studies may be subject to one of various biases, and even if the end result is correct, this does not give any understanding of the components of the risk. On the other hand, epidemiological estimates of risk are still commonsense estimates of the real world risk. There have been many epidemiological studies of waterborne disease of varying quality reported in the scientific literature. Some of these studies are discussed in the chapter on waterborne disease, others are reviewed elsewhere.[3]

4.2.3 Standard Risk Assessment

This is probably the most frequently used risk assessment model in the water industry. It consists of four stages: hazard identification, exposure assessment, dose–response assessment, and risk characterisation (Figure 4.2).[4,5]

The first stage in the process is always hazard identification. Unless the investigator believes he or she has identified a hazard, the rest of the assessment exercise is pointless. A hazard is any agent that has the potential to cause harm to a population and to which the population has been or may be exposed. Thus the risk assessor has to answer two questions:

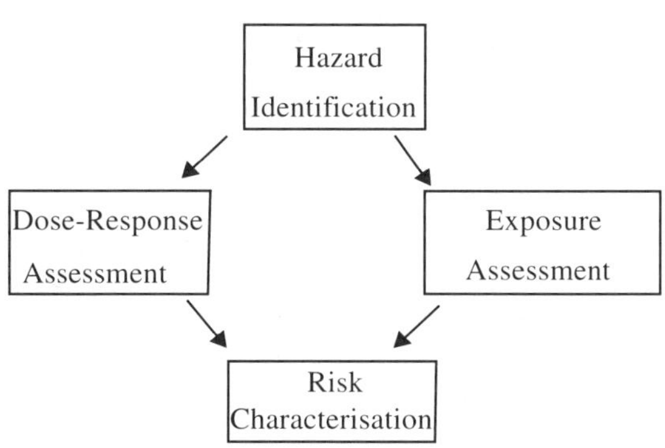

FIGURE 4.2 The risk assessment process.

1. Has the population been exposed to the agent or is it likely to be exposed in the future?
2. Does what is known about the agent suggest that it could cause harm?

Evidence for use in hazard assessment may come from a variety of sources. The results of properly conducted epidemiological studies will provide high quality evidence of hazard. If this is not available, evidence can be obtained from the published results of animal studies. If these are not available, the risk assessor may have to fall back on assumptions from knowledge of the toxicity and pathogenicity of related chemical pathogens. Despite the huge amount of literature on toxicology, epidemiology, and pathogenicity of microbial and chemical agents, the answers to these questions are frequently qualitative rather than quantitative.

Once the hazard has been identified, the next two stages of exposure assessment and dose–response assessment are instigated.

During the exposure assessment phase, the risk assessor has to identify, and wherever possible, quantify the various potential routes of exposure to the agent under investigation. When interested in the water route, the main outcome of this phase will be to identify the actual or probable dose of chemical or pathogen to which the population may be exposed in their drinking water. Clearly, various factors need to be taken into account, such as water consumption behaviour in the exposed population (amount and duration), what is known of the actual concentrations of agent in water supply, and what are the probable concentrations during some sort of adverse event. For chemical agents, the risk assessor is likely to be interested in the lifetime average daily dose (LADD). However, for infectious agents, LADD is not likely to be relevant. Instead, the number of bacteria consumed at a single time will be the best predictor of disease.

The dose–response assessment phase involves assessing the likelihood of ill health for various levels of exposure. Where information is available on the effect of various doses on health, it usually comes from animal studies which may have little or no relevance to humans. Furthermore, these studies usually use doses in greater excess than the levels expected to be found in the environment. Consequently, for real world situations, one has to extrapolate from existing information down to very low doses. There are various approaches to extrapolating down to low doses, but the choice of which method to use is still often fairly subjective. As such, the presumption of adverse health effects at low levels remains, at best, an educated guess.

The final phase is the formal risk characterisation which pulls together all the information gathered in the earlier three phases. Ideally, one should present the data in a standardised numerical form, such as the risk of one extra cancer in a population of 1 million. It is very important that the risk assessor presents the results along with an estimation of the uncertainty of results.

This model of risk assessment has been used extensively for chemical agents. Its use has, however, been more limited for microbial pathogens. Indeed, there are significant additional difficulties in applying the model to microbial pathogens. For example, the susceptibility of individuals in a population varies markedly throughout life and, once infected, whether by the water or another route, people develop

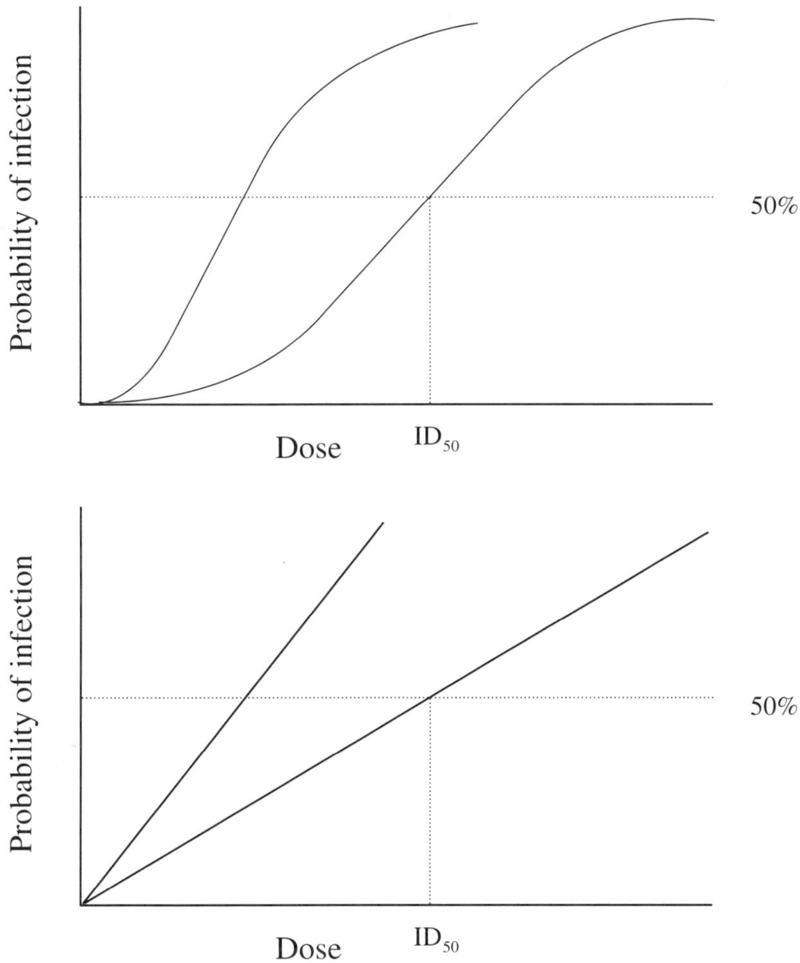

FIGURE 4.3 Effect of model on extrapolating to low doses when infectious dose is not known with certainty.

immunity. Furthermore, an accurate knowledge of the dose–response is further complicated because animal susceptibility to certain pathogens varies considerably from human susceptibility. This is complicated further because there is often considerable variation in infectiousness between different strains of the same microbial species. Figure 4.3 shows how the effect of two different approaches to extrapolating to low doses affects the calculated risk of illness. This figure also shows the impact on different estimations of infectious doses by the two models. It can be seen that the greatest percentage effect on probability of illness would be at low pathogen levels as would be the case in water supplies.

These difficulties have not stopped people from estimating the probability of infection from various doses of microbial pathogens (Table 4.1). Moynihan[7] argues that to get these figures, several unlikely assumptions have to be accepted, namely,

TABLE 4.1
Risk Infection from Exposure to One Organism and Minimum Doses Required to Obtain a 1% Infection within the Community for a Range of Pathogens

Microorganism	Risk Infection from Exposure to a Single Organism	Minimum Dose (No. of Organisms) Required for 1% Infection
Bacteria		
Vibrio cholerae	7×10^{-6}	1428
Salmonella	2.3×10^{-3}	4.3
Shigella	1.0×10^{-3}	10
Campylobacter	7×10^{-3}	1.4
Virus		
Poliovirus 1	1.49×10^{-2}	0.67
Rotavirus	3.1×10^{-1}	0.03
Protozoa		
Giardia lamblia	1.98×10^{-2}	0.5
Entamoeba histolytica	2.8×10^{-1}	0.04

Source: Reprinted from *Water Science Technology*, 24, Rose, J. P. and Gerba, C. P., Use of risk assessment for the development of microbial standards, 29–34, Copyright 1991, with permission from Elsevier Science.

that organisms are randomly distributed, the consumer population is equally susceptible to a single organism exposure, and the exposure is defined as the consumption of 2 l of water per day.[7] The authors would agree with Moynihan's scepticism and would further suggest such estimates are, at best, educated guesses.

An example where the model has been used is in a paper by Rusin and colleagues.[8] This group set out to develop a risk assessment for various microbial pathogens and potential pathogens, *Pseudomonas*, *Acinetobacter*, *Xanthomonas*, *Aeromonas*, *Moraxella*, *Mycobacterium avium*, and *Legionella*. Most of these pathogens rarely, if ever, cause disease in otherwise healthy humans, although all have caused disease in compromised patients. The authors suggested on the basis of their calculations that several of these pathogens could cause community and hospital-acquired disease in humans through drinking water, although case numbers may be very low (less than 1 in 10,000).

More recently, risk assessment models have become considerably more mathematically sophisticated.[9,10] The number of variables in the various mathematical formulae are increasing and the manipulation of these variables has become more complex. For example, in one model, over 20 variables are included, many of which have wide confidence limits.[9] Clearly, the calculated risk has to be so extremely wide that the outputs cannot be accepted as accurate representations of reality.

Nevertheless, these models should not be discounted. Even if the outputs are only estimates of reality with wide confidence intervals, the construction of the models provides a valuable insight into the drivers of risk and their relative importance. Even in the absence of any mathematical or numerical accuracy, the process of thinking through these issues can generate considerable insight into risk.

4.3 CONCLUSIONS

There are varying approaches and models of risk assessment. As the reader will have already discovered, the authors favour the epidemiological approach to risk assessment as being the most realistic. Nevertheless, good epidemiology is difficult to do and can be subject to bias. The mathematical risk assessment models certainly have an important role to play, provided one is not seduced by the apparent accuracy of numerical values.

A potentially significant advance in risk assessment for the water industry would be the introduction of Hazard Analysis and Critical Control Points (HACCP).[11] HACCP has become one of the most useful tools for improving food safety worldwide and has been found invaluable in a range of other industries. Yet despite this success, HACCP has not yet been used to any great extent in the water industry. A more detailed discussion of HACCP will be found elsewhere. This discussion will be restricted to listing the basic principles of HACCP.

1. Conduct a hazard analysis.
2. Determine the critical control points (CCP).
3. Establish critical limits.
4. Establish a system to monitor control of the CCPs.
5. Establish the corrective action to be taken when monitoring indicates that a particular CCP is not under control.
6. Establish procedures for verification to confirm that the HACCP system is working effectively.
7. Establish documentation concerning all procedures and records appropriate to these principles and their applications.

The advantages of HACCP are that it involves people closest to the process, identifies the component of the process that contributes to risk, and identifies mechanisms for risk reduction. Whilst it can truthfully be said that many of these principles have been implemented within the water industry, the authors are not aware of the full process being used. Linking HACCP to existing risk assessment practices would be a major advance in risk assessment.

4.4 REFERENCES

1. Department of the Environment, Transport and the Regions, 1999, The Water Supply (Water Quality) (Amendment) Regulations 1999, *Cryptosporidium* in Water Supplies, HMSO.
2. National Research Council, National Academy of Sciences, 1983, *Risk Assessment in the Federal Government: Managing the Process,* National Academy Press, Washington, D.C.
3. Hunter, P. R., 1997, *Water-Borne Disease: Epidemiology and Ecology*, Wiley, Chichester.
4. Hertz-Picciotto, I., 1995, Environmental risk assessment, in *Introduction to Environmental Epidemiology,* Talbott, E. O. and Craun, G. F., Eds., Lewis Publishers, Boca Raton, FL.

5. U.S. Environmental Protection Agency, 1986, Guidelines for cancer risk assessment, *Fed. Regist.*, 51, 33992.
6. Rose, J. P. and Gerba, C. P., 1991, Use of risk assessment for the development of microbial standards, *Water Sci. Technol.*, 24, 29.
7. Moynihan, M., 1992, Emerging issues for the microbiology of drinking water, M.Sc. thesis, University of Dublin, Dublin, Ireland.
8. Rusin, P. A., Rose, J. B., Haas, C. N., and Gerba, C. P., 1997, Risk assessment of opportunistic bacterial pathogens in drinking water, *Rev. Environ. Contam. Toxicol.*, 152, 57.
9. Eisenberg, J. N., Seto, E. Y. W., Olivieri, A. W., and Spear, R. C., 1996, Quantifying water pathogen risk in an epidemiological framework, *Risk Anal.,* 16, 549.
10. Perz, J. F., Ennever, F. K., and LeBlancq, S. M., 1998, Cryptosporidium in tap water: comparison of predicted risks with observed levels of disease, *Am. J. Epidemiol.,* 147(3), 289.
11. Mortimore, S. and Wallace, C., 1998, *HACCP: A Practical Approach*, Blackie Academic and Professional.

5 Legislation and Water Quality

CONTENTS

5.1 INTRODUCTION

Provision of a safe water supply is essential for human health. There is a basic need for water in sufficient quality and quantity so that it not only sustains life but also allows for continued economic and social development. Water is being used in increasing quantities. It's not just for drinking, but also food production, washing and cleaning, industrial production processes including cooling, recreation, transport, navigation, and irrigation.

As rivers, lakes, and seas, which are used for drinking water abstraction and wastewater discharge, cross national and international boundaries, it is essential that there is cooperation between users of such a vital resource. Pollutants which are put into a river upstream can have health consequences for those abstracting water downstream. There is an increasing awareness of the need for international agreements for the provision of water and management of water resources. The Israelis, for example, supply 55 million m^3 per year of water to Jordan from the Yarmouk river.[1] Such agreements are also necessary for the protection of the quality of water which transcends international boundaries. To facilitate this, there are various international agreements, for example, the ECE Convention on the Protection and Use of Transboundary Watercourses and International Lakes. In June 1999, a draft protocol on water and health was drawn up for submission to the Third Ministerial Conference on Environment Health by the United Nations, the Economic Commission for Europe, and the World Health Organisations Regional Office for Europe.[2] Other international bodies involved with freshwater management and water-related environmental initiatives include the United Nations Environment Programme (UNEP), the Commonwealth, and the Organisation for Economic Co-operation and Development (OECD).

It has been estimated that over 1 billion people,[3] including 100 million Europeans, do not have access to safe drinking water[4] (1997–1998 figures) and that there may be as many as 4 million deaths per year because of limited access to safe drinking water and adequate sanitation.[3] In the developed world, we take for granted that the water coming out of our taps is safe to drink. A recent survey in the U.K. showed that 70% of consumers were satisfied with the quality of water coming out of the tap and that 3 out of 4 drank directly from the tap,[5] and yet we are still affected by outbreaks of waterborne infectious disease. One of the most notable outbreaks was that of cryptosporidiosis in Milwaukee in 1989[6] where it has been estimated that over 400,000 people were affected, 4,000 hospitalised, and up to 50 died.[7] *Cryptosporidia* has also been responsible for outbreaks in Europe.[8] There have also been recent waterborne outbreaks in more of the third world countries, for example, cholera in the Ukraine and Hepatitis A in Estonia.[9]

For water to be acceptable for drinking water purposes, it must be free from contamination which is potentially hazardous to health. Microbiological agents of health related significance include pathogenic bacteria, viruses, protozoa, and parasites. Important bacterial pathogens according to the World Health Organization (WHO) include species of *Campylobacter*, *Salmonella*, *Shigella*, *Vibrio cholerae*, *Yersinia enterolitica*, *Pseudomonas aeruginosa*, *Aeromonas* spp., and pathogenic *Escherichia coli*; viral pathogens include Adenoviruses, Enteroviruses, Hepatitis A, Norwalk virus, Rotavirus, small round structured viruses, enterically transmitted non-A, non-B hepatitis viruses, and Hepatitis E; protozoan pathogens include *Entamoeba histolytica*, *Giardia*, and *Cryptosporidia*, and also a Helminth, *Dracunculus medinensis*.[10]

Water is, therefore, usually treated in some way to render it safe to drink by removal of both chemical and microbial contamination. In the U.K., microbial presence is generally controlled by filtration and halogen-based biocides for disinfection (e.g., hypochlorite) which remove both potential pathogens and nuisance microbes (i.e., those which are not deemed harmful to health but have a detrimental effect on the organoleptic properties of the water supply). Microbial contamination of water supplies is owing primarily to the presence of animal or human faeces which may occur because of an ingress of contaminated water. For example, the contamination can come from untreated sewage entering the distribution system post treatment, animal waste being carried by rain runoff, melting snow, or failure and/or breakdown in the water treatment process.

5.2 LEGISLATION

Legislation is necessary for the protection of the coastal, marine, and inland water environment from pollution, management of water resources, and to maintain and

TABLE 5.1
Examples of Some Directives and Statutory Instruments Relevant to Water Quality

Subject of Legislation	Relevant Directive
Bathing quality of rivers, lakes, and coastal waters	Bathing Water Quality Directive (Council Directive 76/160/EEC concerning the quality of bathing water) and its proposed revision
Wastewater pollution coming from households and industry	Urban Wastewater Treatment Directive (Council Directive 91/271/EEC concerning urban wastewater treatment) and the Integrated Pollution Prevention and Control (IPC) Directive (98C6)05 of 10/1/98
Water pollution	The Nitrates Directive (Council Directive 91/676/EEC concerning the protection of waters against pollution caused by nitrates from agricultural sources)
	Statutory Instrument, 1997 No. 547
	The Control of Pollution (Silage, Slurry, and Agricultural Fuel Oil) (Amendment) Regulations, 1997
	Statutory Instrument, 1999, No. 1006
	The Anti-Pollution Works Regulations, 1999
Drinking water quality	Council Directive 80/778/EC relating to the quality of water intended for human consumption and its revision, Council Directive 98/83/EEC
	The Water Act (c. 15), 1989
	Water Supply (Water Quality) Regulations, 1989
	The Private Water Supplies Regulations, 1991
	The Water Industry Act, 1991
Natural mineral and spring waters and bottled drinking water	Statutory Instrument, 1990, No. 1540, the Natural Mineral, Spring Water and Bottled Drinking Water Regulations, 1999
Cryptosporidia	Statutory Instrument, 1999, No. 1524
	The Water Supply (Water Quality) (Amendment) Regulations, 1999
Materials used in water supplies	Statutory Instrument, 1999, No. 1148
	The Water Supply (Water Fittings) Regulations, 1999

improve the quality of drinking water. The criteria for acceptability of water is based on the World Health Organisation's *Guidelines for Drinking Water Quality*.[10] The guideline values recommended for individual parameters are not mandatory limits but are the basis of national or regional limits included within legislation for acceptable water quality. For example, guidelines are provided in the U.S. within the Safe Drinking Water Act of 1974, subsequent amendments of 1986 and 1996, and National Primary and Secondary Drinking Water Regulations, set and regulated by the Environmental Protection Agency (EPA). In Europe they are provided within the European Union (EU) directives and national legislation (Table 5.1). In Europe, legislation is at both the European level and national level within member states' own water regulations. The member states, however, and not the EU, are responsible for implementing and monitoring once a Directive has been adopted. The European

Commission will, however, take legal action if necessary to ensure that member states comply.

In England and Wales, the regulations pertaining to drinking water quality standards are within The Water Act, 1989 (c. 15) (consolidated in the Water Industry Act, 1991), the Water Supply (Water Quality) Regulations, 1989, and the Private Water Supplies Regulations, 1991, and subsequent amendments and various statutory instruments are enforced by the Drinking Water Inspectorate (DWI) whilst the Department of the Environment (DoE) is responsible for the regulation and control of pollution, the management of water, and flood defenses.

Water quality standards are set to ensure the safety of drinking water supplies through the control or removal from water of constituents known to be hazardous to health. Internationally applied standards for monitoring water quality are essential to allow comparability of analytical data on an international scale.

As regulations are updated, they increasingly require monitoring information to be made available for public review. The Draft Protocol on Water and Health,[2] produced by the United Nations, the Economic Commission for Europe, and the World Health Organisation's Regional Office for Europe[2] state that it is necessary to "establish and to publish national and/or local targets for the standards and levels of performance that need to be achieved or maintained for a high level of protection against water-related disease." Such targets should include a reduction in the scale of outbreaks and incidents of water-related disease. This document requires that not only is surveillance necessary to identify outbreaks, potential outbreaks, and/or incidents which affect public health, but also that information is disseminated to the public.

This requirement to inform the consumer of water quality is included in European legislation within the latest drinking water directive, and can be seen in the Council Directive 98/83/EC, adopted by Council on November 3, 1998, which said, "on the quality of water intended for human consumption," and came into force on December 25, 1998, replacing Council Directive 80/778/EC, due to be repealed in 2003. Member states have 2 years to transpose it into national regulation and 5 years, until 2003, to ensure that they comply with the new standards, except for those for bromate, lead, and trihalomethanes, which have 10, 15, and 10 years, respectively.

Materials used in distribution systems and within domestic properties also affect the quality of water provided to the consumer at the tap. In the U.S., the materials used in water systems are covered by American Standards Institute (ANSI)/NSF International Standard 61, but compliance with this standard in some states is voluntary and not a mandatory part of federal law[11] although now more than 40 states do require compliance with this standard.[12] A major change in the new Directive includes the *Tapwater Directive*. The point of use is now the point of compliance, with an obligation to inform the consumer of measures that they can take to comply with the requirements of the Directive where the components of the system are the cause of the noncompliance. A recent Statutory Instrument, No. 1148, The Water Supply (Water Fittings) Regulations, 1999, has been included in the U.K. legislation which covers the installation and design of fittings to prevent contamination of the water supply.

The new Directive has reduced the measuring parameters from 66 in 80/778/EC to 48 (50 for bottled waters) and reviewed the parametric values, including a reduction in the lead and copper content and more stringent pesticide and organic chemical levels. Test methods are now referenced to ISO/CEN standards which have implications for national test methods and will be discussed next.

Directive 98/83/EC does not apply to natural mineral waters or waters which are intended for medicinal use. The Natural Mineral, Spring, and Bottled Drinking Water Regulations, 1999, the International Standards for Mineral Waters, and the Codex Alimentarius Commission (CAC) adopted a worldwide Standard for Natural Mineral Waters in 1997. As well as establishing limits for chemical constituents of the water and microbiological parameters, it specifies a code of practice for safe handling and processing.[13]

Regulations need to be updated as new or previously unrecognised waterborne vectors of disease are discovered. A recent example of this in the U.K. is the The Water Supply (Water Quality) (Amendment) Regulations, 1999, supported by Guidance on Assessing the Risk of *Cryptosporidium* Oocysts in Water Supplies and Standard Operating Protocol for Monitoring of *Cryptosporidium* in Water Supplies. Cryptosporidiosis, a parasite of animals and humans, may pass through water treatment plants in sufficient numbers to cause gastrointestinal disease which can have serious consequences for high risk members of the population, particularly if they are immunocompromised (e.g., AIDS patients). These new regulations which came into force July 1999 make it a criminal offense for water companies who do not adequately treat their water supply for *Cryptosporidium.* Water companies are required to carry out assessments to identify supplies which carry a risk of *Cryptosporidium* infection and, if a risk is identified, to continuously monitor for its presence.

5.3 MONITORING WATER QUALITY

Most of the fresh water we use is abstracted from surface waters and from rivers. We need to remain aware, however, that all water which has been used, whether by households, agriculture, or industry, will eventually find its way back into the environment and the watershed and, therefore, must be monitored to ensure it is of potable quality both chemically and microbiologically.

To protect public health and ensure compliance with regulations, routine monitoring of water is carried out to assess the effectiveness of disinfection and detect the ingress of possible faecal contamination, whether used for drinking, food preparation, or when there could be human contact (e.g., washing and bathing). Monitoring is required under legislation in both Europe and the U.S. In the U.K. the DWI 1999 annual report showed that 99.75% of approximately 3 million tests performed met the relevant water quality standards with an expenditure of around £2.1 million (DETR figures).[14]

Whilst it is possible to detect a range of potential human pathogens in water including bacteria, protozoa, parasites, and viruses, this is not a practical proposition for day-to-day monitoring and would only be normally undertaken if water were to be implicated as a source of an outbreak of enteric disease. Many of the methods for the detection of pathogens are time consuming and expensive, and as pathogens are likely to be present only in low numbers, they are not easily detected even in untreated surface waters (usually a few per litre) and even less in treated waters.[15] Many waterborne enteric pathogens are not bacterial and, therefore, are not detectable by traditional culture techniques. Molecular and immunological methods, including methods for detection of viruses and protozoa, are being developed and will also be capable of giving information on the virulence of the target organism.

5.4 INDICATOR ORGANISMS

Microbiological monitoring needs to be rapid and sensitive; bacteria commonly found in the animal and human gut indicate the presence of faecal contamination and are used as bacterial indicators of faecal pollution. Coliforms from human and animal waste may be found in drinking water which has not been properly disinfected. To be useful as an indicator of faecal pollution, indicator organisms should not be able to grow in natural water and their persistence and, therefore, susceptibility to water treatment within water should be similar to the pathogens that cause waterborne enteric disease.

The presence of indicator organism coliforms and *E. coli* suggests that intestinal pathogens could also be present and the supply could, therefore, pose a threat to health. However, the converse, that is, the absence of indicator organisms, does not necessarily preclude the presence of pathogens.[16] It is not possible from a sample which is positive for indicator organisms to predict the number of pathogens likely to be present. Neither does a single negative sample suggest that the supply is safe. A regular programme of monitoring should be implemented using the relatively simple tests developed to detect indicator organisms. Coliform bacteria and *E. coli* are the most common indicator organisms used in routine monitoring of drinking water quality.

5.4.1 COLIFORMS

Coliforms are members of the *Enterobacteriaceae*, which may not be of animal origin, may occur naturally in soils and on decaying plant material, and may multiply in water rich in organic nutrients.[10] As a result, coliforms are, therefore, of limited use as indicators of faecal pollution, but to date are still accepted as such. Coliforms are capable of growth at 37°C, and possess β-galactosidase. This definition also includes anaerogenic strains. Coliforms are, by definition, oxidase negative and catalase positive. The following genera are commonly isolated: *Citrobacter*, *Enterobacter*, *Escherichia*, *Hafnia*, *Klebsiella*, *Serratia*, and *Yersinia*.

Some bacteria found in water may conform to the definition of coliform organisms in most respects, but are able to produce gas from lactose only at temperatures below 37°C. *Aeromonas* species, which occur naturally in water, have an optimum growth temperature in the range 30 to 35°C but may, nevertheless, produce acid, with or without gas, from lactose at 37°C. They are of uncertain public health significance but have been implicated in causing gastrointestinal (GI) disease. They are distinguishable from the coliform group by a positive oxidase reaction.

5.4.2 *Escherichia coli*

Escherichia coli (*E. coli*) is universally found inhabiting the guts of warm-blooded animals, including humans, and is found in large numbers, commonly, approximately equal to 10^9 per gram. Its presence within water systems suggests that there has been a faecal contamination incident of either human, animal, or bird origin and, therefore, the presence of more pathogenic organisms is possible. For water-monitoring purposes, the definition of *E. coli* is based on the fermentation of lactose or mannitol at 44°C, the production of acid within 24 hours, and the production of indole from tryptophan at 44°C. Most strains produce β-glucuronidase. Though some strains may express these characteristics at 37°C and not at 44°C, they could also be *E. coli.*

Confirmation of the presumptive result is by characterisation of coliforms and *E. coli.*[17] Coliforms are identified by the presence of typical colony morphology, production of acid from lactose, and lack of cytochrome oxidase. In addition, the presence of *E. coli* can be demonstrated by the production of indole in tryptone water[18] or by use of a commercial identification kit.

5.4.3 Enterococci (Faecal Streptococci)

Enterococci are Gram-positive cocci, oxidase and catalase positive, possess the Lancefield Group D antigen, and can grow in the presence of azide at concentrations which inhibit the growth of coliforms. They are almost always present in the human gut in large numbers, though some species may also be isolated from animals and are almost always outnumbered by *E. coli*. They generally do not multiply in natural water systems, are more resistant to drying, and persist longer than coliforms and *E. coli*. They are useful as a secondary indicator of treatment efficiency, e.g., where there are coliforms present but not *E. coli*, to confirm the presence of faecal pollution.[16] They have also been shown to be better than coliforms or *E. coli* as indicators of the risk to health from bathing water.

5.4.4 Sulphite Reducing Clostridia and *Clostridium perfringens*

Clostridium perfringens are spore forming, Gram-positive, obligate anaerobes normally present in faeces, and reduce sulphite to sulphide. The spores are relatively more resistant to chlorine than the other indicators mentioned and will, therefore, survive longer in water. Their presence in the absence of the more chlorine-sensitive coliforms, *E. coli* and enterococci, suggests that there has been a more temporally distant faecal contamination event. For monitoring purposes, there should be 0 per 100 ml. Because of their chlorine resistance, it has been suggested that the presence of the spores of *C. perfringens* may be useful as an indicator of *Cryptosporidium*.

5.4.5 Heterotrophic Colony Counts

Heterotrophs are generally not believed to be the cause of disease in humans but their presence may give an indication that control within a distribution system is not satisfactory. Colony counts are only useful as a parameter if they are used continuously to monitor a system. A sudden increase over the numbers normally found in a particular system or a sustained increase over a period of time is suggestive of a failure in disinfection or colonisation of the system, respectively.

5.5 METHODS USED TO DETECT INDICATOR ORGANISMS

Standardisation of methods is essential to reduce the variation in results. The final result may vary according to the time taken and maintenance of sample conditions between sampling and testing, growth medium used, incubation conditions, and procedures used for resuscitation of stressed and/or damaged microorganisms following disinfection. Comparison of national and international water quality cannot be made unless methods are internationally comparable.

International standards methods exist, for example, those produced by the International Organisation for Standardisation (ISO methods) and Comité Européan de Normalisation [European Committee for Normalisation (CEN)]. These are now referenced within the new EU drinking water directive, where ISO 9308-1 is specified as the standard test method for coliforms and *E. coli*. This ISO is under revision and the draft ISO specifies the use of lactose TTC agar (the French standard method). The Directive also specifies m-cp agar for *Clostridium perfringens* (the Dutch standard method). Neither of these methods are routinely used in the U.K. In countries where the methods for drinking water analysis are not those specified within the Directive (e.g., within the U.K.), alternative methods must be validated to show that they are of "equivalent or better" performance when compared to the method specified. In the U.K., standard methods based on those described in Report 71[16] are generally in use, but it will be necessary to validate their performance against the methods specified in the new Drinking Water Directive. For comparison of results, testing laboratories should also be of a comparable standard in their performance of the tests. In the U.K., the laboratories must be accredited by the U.K. Accreditation Service (UKAS) or the Drinking Water Inspectorate for the relevant methods.

Various methods are available to detect coliforms and *E. coli,* which for drinking water purposes should be absent from 100 ml of the sample. A multiple tube method utilises a series of tubes of different volumes of sample added to a double-strength medium, for example, Minerals Modified Glutamate Medium (MMGM),[16] to give the most probable number (MPN) of indicator bacteria in the sample. The presence of acid, denoted by a change of colour to yellow, indicates a positive result[19] and is particularly suited for testing waters of high turbidity. A simplified version which does not give a quantitative result but is based solely on the presence/absence (PA) test of indicator bacteria, incorporates one large volume of medium (100 ml) instead of the series and may be obtained in a commercially available kit form.

Such kits are acceptable, for example, when used for testing water leaving a treatment works where daily sampling is required. PA kits may also be useful for testing nonroutine samples, for example, after a distribution system repair.

A commercial variation of the MPN method is available and uses IDEXX Colilert 18® in Quantitrays™ to detect the presence of the enzymes β-galactosidase by a colour change to yellow and β-glucuronidase by fluorescence under long wavelength UV light. It gives the simultaneous detection and confirmation within 24 hours of coliforms and *E. coli*, respectively. Membrane filtration is more often used as the method of choice. Generally, the sample (usually 100 ml) is concentrated onto a membrane by filtration and the membrane is then placed onto a suitable selective medium or pad soaked in a selective broth. For coliforms and *E. coli* in the U.K., the medium used is usually membrane lauryl sulphate broth (MLSB).[16] Presumptive counts, that is the number of organisms that are of typical appearance and/or produce a typical result for the target organism on the primary detection medium, are noted.

On MLSB, presumptive total coliform bacteria are organisms which produce acid from lactose, form all shades and sizes of yellow colonies on membranes after incubation for 4 hours at 30°C followed by 14 hours at 37°C, and are oxidase negative. Isolates which produce acid from lactose after incubation for 4 hours at 30°C followed by 14 hours at 44°C are presumptive *E. coli.* A confirmed result, which for *E. coli* is the production of indole from tryptophan, having a negative oxidase result is usually available after 48 hours.

5.6 POSITIVE RESULTS

Potable water should not contain pathogens and, therefore, the counts of indicator organisms should be 0 per 100 ml. In most cases where there has been an epidemic of waterborne disease, it has been found subsequently that there has been a lapse or failure in the water treatment process.[10] Following a presumptive positive result from a treated water sample, whether from the treatment works, service reservoirs, or consumer taps, immediate action should be taken. Further samples should be taken from the same site immediately for analysis by a quantitative method in a microbiology laboratory. Samples may also be taken from related locations and examinations extended to include enterococci (faecal streptococci) and *Clostridia.* If a negative result is obtained after 18 to 24 hours, the test should be incubated for a further 24 hours to give a final result.[20] Concurrently, confirmation of the presumptive result should be carried out and checks made to ensure that the sample has been taken correctly. Checks should also be carried out to ensure that treatment has been successful and uninterrupted and there has been no failures in the integrity of the distribution system.

5.7 CONCLUSION

As the population increases, the demand on water resources is also going to increase. International agreements for sharing water resource management will become more important and it is, therefore, necessary that there should be international standards of water quality and water quality monitoring. With the predicted increases in population and global warming, availability of good quality water may well become a matter of international dispute. It is, therefore, vitally important that management of wastewater and pollution water quality is more pro-active than re-active in nature and that regulatory bodies move toward a rolling program of legislation update with an increased openness to public scrutiny.

5.8 ACKNOWLEDGMENTS

We would like to thank Sue Surman and John Lee for hlep in the compiling of this chapter.

5.9 REFERENCES

1. *Water Law and Policy Programme,* Newsletter, University of Dundee, 1999.
2. Draft Protocol on Water and Health, 1999, to the 1992 Convention on the Protection and Use of Transboundary Watercourses and International Lakes, for submission to the Third Ministerial Conference on Environment and Health, London, June 1999.
3. **Anon.,** 1998, NSF, PAHO and WHO Focus World on Universal Access to Safe Drinking Water, NSF Report on NSF, WHO, PAHO Symposium on Safe Drinking Water, Washington, D.C.
4. **Anon.,** 1997, Will the tap run dry?, Economic Commission for Europe Publication, ECE/ENV/97/5.
5. **Anon.,** 1999, *Periodic Review 1999,* Consumer Consultation by The Drinking Water, summary of report on research findings.
6. Mackenzie, W. R., Hoxie, N. J., Proctor, M. E., Gradus, M. S., Blair, K. A., Peterson, D. E., Kazierczak, J. J., Addiss, D. G., Fox, K. R., Rose, J. B., and Davis, J. P., 1994, A massive outbreak in Milwaukee of cryptosporidiosis infection transmitted through a public water supply, *New Eng. J. Med.,* 331, 161.
7. **Anon.,** 1999, Water on Tap, A Consumer's Guide to the Nation's Drinking Water, EPA publication.
8. Richardson, A. J., Frankenberg, R. A., Bach, A. C., Selkno, J. B., Colbourne, J. S., Parsons, J. W., and Mayon-White, R. T., 1991, An outbreak of waterborne cryptosporidiosis in Swindon and Oxfordshire, *Epidemiol. Infect.,* 21, 126.
9. **Anon.,** 1997, Water and Health: UN/ECE fights for clean drinking water, ECE/ENV/97/11, Economic Commission for Europe Publication.
10. *Guidelines for Drinking Water Quality,* 1993, Vol. 1, Recommendations, World Health Organisation.
11. **Anon.,** 1998, *Waterworks,* Vol. 2, No. 1, March 1998, NSF International.
12. **Anon.,** 1999, *Waterworks,* Summer 1999, NSF International.
13. Recommended International Code of Hygienic Practice for the Collecting, Processing and Marketing of Natural Mineral Waters, CAC/RCP 33-1985, Codex Alimentarius, Vol. 11, 1994, FAO, Rome.
14. Department of the Environment Transport and the Regions DETR Annual Report, 1999, the goverment's expenditure plans, 1999–2000 and 2001–2002.
15. Havelaar, A. H., 1998, Emerging microbiological concerns, in *Proc. Drinking Water Res.* 2000, University of Warwick, Drinking Water Inspectorate.

16. **Anon.,** 1994, The Microbiology of Drinking Water 1994, Part 1: Drinking Water, Report on Public Health and Medical Subjects, No. 71, Methods for the Examination of Waters and Associated Materials, HMSO.
17. Cowan, S. T., 1993, *Cowan and Steels' Manual for the Identification of Medical Bacteria,* 3rd ed., Barrow, G. I. and Feltham, R. K. A., Eds., Cambridge University Press, London.
18. Kovacs, N., 1928, Eine vereinfachte Methods zum Nachweis der Indolbildung durch Bakterien, *Z. Immunit. Exp. Ther.,* 55, 311.
19. Public Health Laboratory Service, 1969, A minerals modified glutamate medium for the enumeration of coliform organisms in water, by the Public Health Laboratory Service Standing Committee on the Bacteriological Examination of Water Supplies, *J. Hyg.,* 67, 367.
20. **Anon.,** 1990, *Guidance on Safeguarding the Quality of Public Water Supplies,* HMSO.

6 Biofilm Development in General

CONTENTS

6.1 INTRODUCTION

Biofilms have been cited in the literature for a number of years, often being defined as, "cells immobilized at a substratum and frequently embedded in an organic polymer matrix of microbial origin."[1,2] Whilst this definition of a biofilm is acceptably portrayed as the universally acknowledged biofilm model, slight reclassification has taken place. This occurred in 1995 with the redefinition of biofilms being "matrix-enclosed bacterial populations adherent to each other and/or to surfaces or interfaces."[3]

Despite ongoing discussions on the so-called biofilm model, the enormous diversity of biofilms evident today suggests that strict phraseology for a constantly changing dynamic ecosystem is not possible. As Stoodley et al.[4] have suggested, it may not seem necessary to "restrict a biofilm model to certain structural constraints but instead look for common features or basic building blocks of biofilms." With this in mind, it seems plausible to suggest that biofilms form different structures and are composed of different microbial consortia dictated by biological and environmental parameters which can quickly respond and adapt both phenotypically, genetically (possibly), and structurally to constantly changing internal and external conditions.

Consequently, it seems illogical to suggest that a true biofilm model system can be achieved so that it can be applied to every ecological, industrial, and medical situation. Therefore, the definition of a biofilm has to be kept generalised and could be redefined as, "microbial cells, attached to a substratum, and immobilised in a three-dimensional matrix of extracellular polymers enabling the formation of an independent functioning ecosystem, homeostatically regulated."

6.2 WHY A BIOFILM?

Within nature, the human body, and industrial surroundings, it is now widely accepted that the majority of bacteria exist, not in a free-floating planktonic state but attached to surfaces within biofilms. As a consequence of this phenomena, there must be, without being too anthropomorphic, advantages to microbial populations in the attached sessile state, particularly, as it is well documented, where at surfaces, bacteria are known to confer a number of advantages not evident when compared to their planktonic counterparts.

The advantage of sessile growth as opposed to the planktonic state include

- The expression of different genes (beneficial genes).[5]
- Alterations in colony morphology[6]—some *Pseudomonas* sp. form filamentous cells when grown as a biofilm as opposed to rod-shaped cells when grown in a liquid culture.
- Different growth rates which are known to aid antimicrobial resistance.[7]
- Larger production of extracellular polymers (possibly aiding antimicrobial resistance).[8]
- Enhanced access to nutrients.[9]
- Close proximity to cells with which they may be in mutalistic or synergistic association.
- Protection to a high degree from various antimicrobial mechanisms, that is, biocide, antibiotics, antibodies, and predators.[10,11]

The substratum surface to which the biofilm is attached, also provides protection and offers resident bacteria a nutritional advantage over their planktonic counterparts so that surfaces are the major site of microbial activity,[12] particularly in water distribution systems.[13] Many aquatic bacteria depend on attachment to surfaces for survival, with sessile cells growing and dividing at nutrient concentrations too low to permit growth in the planktonic phase.[14]

The sessile mode of growth also seems to be important for both the survival and reproductive success of microorganisms. Biofilms, particularly, act as reservoirs of bacterial species, sites of specific limited niches, and protective sites from competition and predators.

The incorporation of bacteria within a biofilm seems to suggest a survival strategy of bacteria. This adaptive strategy, partially if not wholly, relates to both the physical and chemical nature of the environment to which the sessile microbes are associated. Whilst this is true, what must also be considered is that bacterial communities have the capabilities to alter the environment to which they are associated. This would have fundamental effects on the sessile bacterial communities and viability and sustainability of the biofilms associated with a surface.

Whilst surface adhesion and colonisation differ substantially from species to species, there are a number of fundamental processes common to all sessile bacteria. For example, all bacteria must

- Attach to a substratum or other bacteria.
- Have the ability to utilise available resources for growth and reproduction.
- Have the ability to redistribute to different areas if local conditions become unfavourable.

With the constantly changing conditions within a biofilm, sessile bacteria must be able to survive these changes and adapt over time. In order for this to be achievable, bacteria must remain simple, diverse, and metabolically adaptable.

The dynamics of biofilms make the existence of a pure culture biofilm within both natural and industrial situations an unrealistic survival strategy and a system not often encountered, if at all. This, however, is not necessarily true of medical biofilms where surfaces are often associated with biofilms containing monocultures of either *Pseudomonas aeruginosa* or *Staphylococcus aureus*.

6.3 MECHANISMS BEING USED TO STUDY BIOFILMS

With the use of the electron microscope, researchers have identified the presence of microorganisms enclosed in an extracellular polymeric substance (EPS) which are associated with surfaces.[15-17] Biofilms and bacterial adhesion have also been studied with the use of scanning confocal laser microscopy (SCLM), microbalance applications, microelectrode analysis, high-resolution video microscopy, atomic force microscopy, and scanning electron microscopy. Systems used to study biofilms are discussed in Chapter 9.

6.4 STAGES IN THE FORMATION OF BIOFILMS

Bacteria generally range in size from 0.05 (nanobacteria) to 4 μm in length or diameter, with slow-growing and starved cells dominating at the smaller end of the range and fast-growing cells, especially in nutrient rich environments, at the larger end. Bacteria commonly bear a negative charge[18] with the initial interactions between bacteria and surfaces being considered in terms of the colloidal behaviour.[19] However, the fact that bacteria are living entities and capable of changing themselves and their environment through active metabolism and biosynthesis must not be overlooked.[18]

The process of biofilm formation is now considered to be a complex process, but generally, it can be recognised as consisting of five stages. These include (Figure 6.1)

1. Development of a surface-conditioning film.
2. Those events which bring the organisms into the close proximity with the surface.

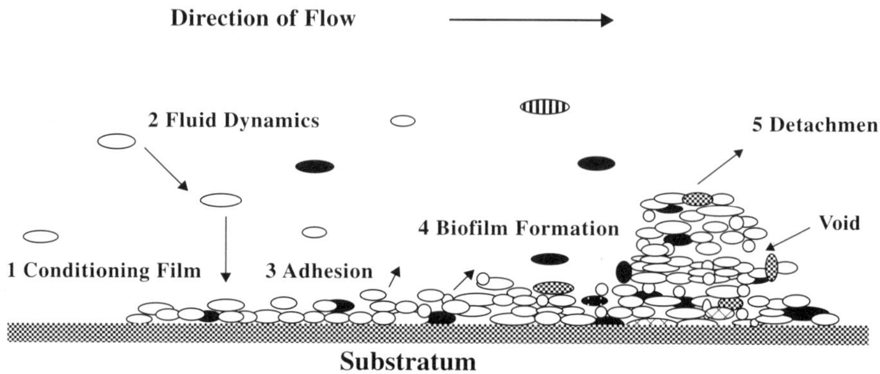

FIGURE 6.1 Diagram to show biofilm formation.

3. Adhesion (reversible and irreversible adhesion of microbes to the conditioned surface).
4. Growth and division of the organisms with the colonisation of the surface, microcolony formation and biofilm formation.
5. Detachment.

Each of these processes will be considered in turn.

6.4.1 Development of the Conditioning Film

Marshall[20] described a surface evident in a flowing system as a "relatively nutrient-rich haven in an otherwise low nutrient environment." This quote suggests that clean unexposed surfaces when evident in either natural or *in vitro* solutions become conditioned with nutrients. Whether these molecules which condition the surface function as microbial nutrients is largely unknown. It does, however, seem to be generally accepted that a clean surface which first makes contact with a bathing fluid must have organic substances and microbial cells transported to the surface before biofilm development can begin. Despite the presence of a conditioning organic film, there has been some discussion as to whether or not it is a prerequisite for bacterial attachment. This problem is difficult to resolve because it is unlikely that any surface is absorbate free before microbial attachment occurs. Adsorption begins immediately on immersion of an unexposed, clean surface to a bathing liquid. Studies that have been carried out indicate that conditioning of surfaces occurs after being exposed to a bathing fluid for 15 min.[21,22] with the thickness of these initial films being calculated at between 30 and 80 nm.[23]

The conditioning film in nature seems, therefore, to play a major role in modifying the extent of bacterial adhesion to immersed surfaces. This seems a plausible statement, particularly because the nature of the adsorbed layer depends very much upon the environment to which the surface is exposed.

Before a surface is exposed to a bathing fluid, it is either negatively or positively charged. After exposure to bathing fluid, surfaces acquire a negative charge owing to the adsorption of macromolecules such as humic acids, low molecular weight, and hydrophobic molecules, which condition the newly exposed surface.[24,25] In aquatic or terrestrial environments, the major components of the conditioning film are likely to be organic. Particularly in these situations, the conditioning layer has been shown to consist of complex polysaccharides, glycoproteins, and humic compounds.[26]

Research with the Fourier-Transformed Infrared spectroscopy (FTIR), multiple Attenuated Internal Reflectance Infrared spectroscopy (MAIR-IR), and Infrared spectroscopy (IR) has also found evidence that the conditioning film contains glycoproteins, proteins, and humic substances.[27-29] The way in which these molecules interfere and amplify the adhesion process remains unclear. However, it is generally acknowledged that these conditioning chemicals can interact with surface appendages evident on bacterial species. These include the pili, fimbriae, glycocalyx, and EPS.[30-33] It is well documented that certain surface appendages are capable of extending through the energy barrier evident during the adhesion process, allowing for some contact to be made with the conditioned surface film.

The conditioning film is regarded as both chaotic and dynamic with no indication of it being static, with adsorbed molecules on surfaces desorbing or disappearing with exposure time. However, the conditioning film is generally observed or presumed to be uniform in both composition and coverage, but to date, research suggests that there appears to be little conclusive evidence to suggest that the spatial distribution of the conditioning film is uniform so that an uneven and heterogeneous development is possible. This, ultimately, will affect both the microbiological composition and development of the biofilm.

Overall, in view of the available literature, it has been suggested that the roles of the conditioning film in the process of bacterial adhesion include[26]

- Modifying physico-chemical properties of the substratum.
- Acting as a concentrated nutrient source.
- Suppression of release of toxic metal ions.
- Adsorption and detoxification of dissolved inhibitory substances.
- Supply of required metal trace elements.

It may also act as a triggerable sloughing mechanism or suppress/inhibit the adhesion of bacteria induced by surface polymers. However, this needs further investigation to warrant validity.

6.4.2 Transport Mechanisms Involved in Adhesion of Microorganisms

In very dilute solutions containing low concentrations of microbial cells and nutrients, transport of microbial cells to the substratum may be the rate-controlling step in biofilm accumulation and, therefore, fundamental to the understanding of biofilm formation.

The transport of microbial cells and nutrients to a surface can be explained by a number of well-known fluid dynamic processes. These include

- Mass transport, which is influenced strongly by the mixing in the bulk fluid and being related to water flow rate, that is, laminar or turbulent.
- Thermal effects (Brownian motion, molecular diffusion).
- Gravity effects (differential settling, sedimentation).[34]

Within pipes transporting potable water, two main flow conditions are known to be evident, namely laminar and turbulent flow.[35] Generally, laminar flow can be characterised as having parallel smooth flow patterns with little or no lateral mixing with the fastest flow in the centre (Figure 6.2).[36,37] This type of flow is known to occur in the bloodstream and urinary system where microorganisms and nutrients are considered to keep a straight path and remain in a stabilised position dictated by the flow rate.[37]

Turbulent flow, however, is flow which is random and chaotic allowing for bacteria and nutrients to be mixed and transported nearer to the surface than in laminar flow (Figure 6.3). Because this type of flow is complex and ultimately difficult to predict, most research in the area of adhesion and transport mechanisms has been with laminar flow.[22]

When a fluid first enters a pipe, it has almost uniform velocity. As the fluid moves along the pipe, viscous effects cause it to stick to the pipe wall.[35] Hence fluid moving near the centre of the pipe is more rapid than fluid moving near the wall

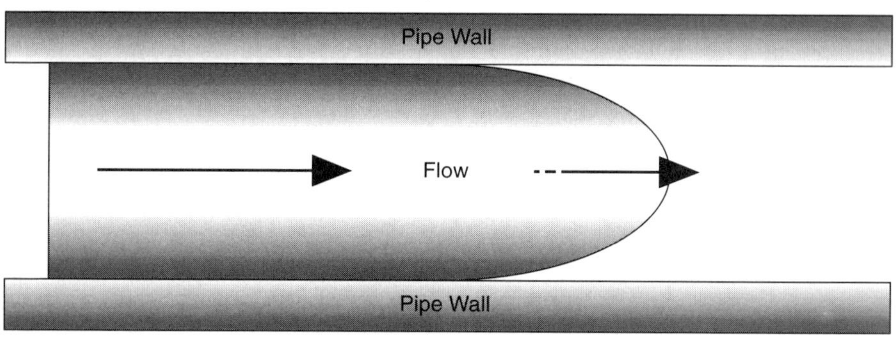

FIGURE 6.2 Diagrammatic representation of laminar flow through a pipe system.

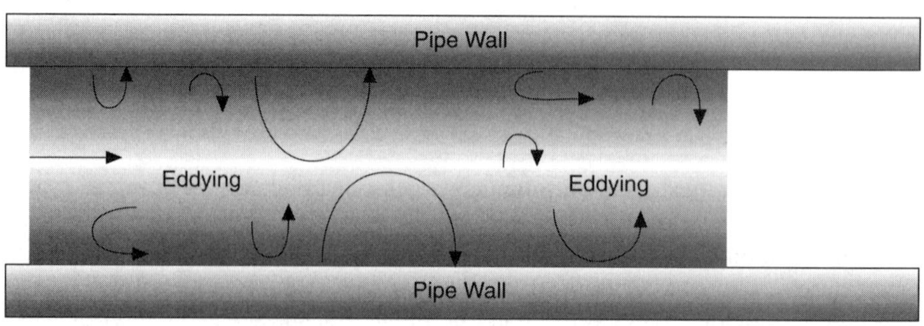

FIGURE 6.3 Diagrammatic representation of turbulent flow through a pipe system.

owing to the drag caused by the viscosity.[38] Owing to this effect, differences exist in velocity profiles between laminar and turbulent flow. In both laminar and turbulent flow regimes, the fluid next to the surface of the pipe wall begins to form a boundary layer in which the viscous forces are more important than the acceleration or inertia forces.[39] As a result of these viscous forces, the fluid in the boundary layer is separated from the fluid outside the boundary layer. In laminar flow, the fluid in contact with the pipe has zero velocity resulting in the development of a velocity gradient between the fluid in the free stream and the pipe surface. When the boundary layer becomes turbulent, the flow immediately next to the solid surface is not. Therefore, a thin layer (1 μm) exists adjacent to the solid surface in which the flow has negligible fluctuations in velocity.[39] This area is called the laminar or viscous sublayer.[38,40]

In laminar flow, the boundary layer takes up the whole of the pipe with the flow close to the pipe surface being much slower. This area has been referred to as the stagnant layer owing to mass transfer limitations. This would suggest that biofilm formation/development within laminar flow is subjected to a number of limitations, particularly that of nutrient supply. The lack of mixing and slow velocity near the surface depletes nutrient supplies to the biofilm substantially.[38] Also, the possibility of toxic waste product buildup in the vicinity of the biofilm should not be ruled out because this would also affect the biofilm development, often leading to biofilm detachment.[41]

However, turbulent flow, a situation more relevant to water distribution systems, also has effects on biofilm development particularly that of organism deposition and nutrient delivery.[36] In turbulent flow, the boundary layer remains very close to the pipe surface and is considered to be where laminar flow predominates and most of the resistance to mass transfer occurs.[22] The boundary layer does not fill the radius of the pipe as in laminar flow. The sublayer is constantly penetrated by turbulent fluctuations and bursts. This is one way bacteria are thought to be transported to the pipe surface.

Eddying currents (random and unpredictable flow) are evident in turbulent flow which cause up and downsweep forces which extend from the bulk flow of fluid and penetrate all the way to the pipe surface. This helps to propel bacteria to within a short distance of the surface, enabling an increased chance of adhesion. If bacteria are travelling faster than the fluid in the region of the wall, a lift force directs the bacteria toward the wall.[34] In the boundary layer, the bacteria encounter significant frictional drag forces which gradually slows down a bacterium as it approaches the surface. There is also a fluid drainage force resulting from the resistance a bacterium encounters near the wall. This is owing to the pressure in the draining fluid film between the wall and approaching bacterial surface. Aside from eddy currents, another mechanism for directing particles through the boundary layer to the pipe wall is turbulent downsweeps. These spontaneous bursts of turbulence penetrate the viscous sublayer and provide a significant fluid mechanical force to direct the bacteria to the solid surface. This provides the means of transporting bacteria from the bulk phase to the vicinity of the wall.

Overall, fluid dynamic forces serve to disperse microorganisms throughout a liquid phase but seem also to concentrate the suspended organisms in the proximity of the viscous sublayer. Research on the structure of the viscous sublayer in turbulent flow indicates that downsweeps of fluid from the turbulent core penetrate all the way to the wall[42] and may transport particles from the bulk fluid all the way to the wall. Aside from lift, this is the only fluid mechanism force directing the particle to the wall. This seems to be a very important process in turbulent flow systems. Within flowing systems, other mechanisms aid in the transport and adhesion of cells to surfaces. These are a part of Brownian diffusion, which has little effect on the movement of bacteria in aquatic systems and thermal gradients, which may contribute to the transport of microbial cells to or away from the surface.[43]

Another parameter which may influence transport and attachment of microorganisms to a surface is the chemical environment in which a bacterium exists. These adhered chemicals would influence the direction of taxis[44] with chemicals that elicit positive chemotactic responses. This would enhance the rate of bacterial attachment to artificial surfaces and chemicals, which cause negative chemotactic responses leading to active avoidance of certain regions.[45] The negative chemotactic response of certain bacteria to sublethal concentrations of toxins has been shown to take precedence even when higher concentrations of nutrients or other chemicals, which usually cause a positive chemotactic response, are present.

In static or quiescent environments, adhesion is aided by a number of factors including Brownian diffusion, gravity, and motility.[27] Generally, it is motility which increases the chances of bacterial adhesion.[46,47] This is possibly owing to enough potential energy available to overcome any repulsive forces known to operate between the bacterial surface and the substratum in question. To reinforce this supposition, it is generally found that the reduction in motility as a result of culture age leads to a reduction of adsorption.[46] Other mechanisms are also known to be evident as factors governing surface colonisation and include gravitational cell sedimentation, often only of relevance in flowing systems when co-aggregation is evident.[48]

Fluid dynamic forces are also known to affect the structure of the developing and developed biofilm. Turbulence is known to increase attachment of microbial cells to a surface, but if a biofilm becomes too thick, detachment is known to occur. This occurs when the biofilm extends past the boundary layer. It is not until the biofilms protrude through the sublayer that the frictional resistance increases.[49] This, ultimately, would have an effect on the flow in the pipe effectively causing a decrease in flow rate.[50] If a biofilm protrudes through the viscous sublayer, there is increased turbulence in the biofilm vicinity and, therefore, an increased rate of erosion, sloughing, and abrasion.

6.4.3 Reversible and Irreversible Adhesion

After conditioning of the substratum and transport of bacteria into the boundary layer, adhesion may take place. Studies carried out on bacterial adhesion, first introduced by Zobell in 1943,[51] suggest that adhesion consists of a two-step sequence comprising: reversible adhesion and irreversible adhesion.

The process of adhesion was later redefined by Marshall et al.[27] in 1971 as, "reversible and irreversible sorption." Reversible adhesion is referred to as an initial weak attachment of microbial cells to a surface—cells attached in this way still exhibit Brownian motion and can easily be removed by mild rinsing.[52] Conversely, irreversible adhesion aided by extracellular polymeric substances establishes a permanent bonding of the microorganisms with the surface requiring mechanical or chemical treatment for removal.

Microbial adhesion has been described in the literature in terms of DLVO theory developed and named for Derjaguin and Landau[53] and Verwey and Overbeek[54] to explain the stability of lyophobic colloids, representative of bacterial cells, and the surface free/hydrophobicity theory.

The DLVO theory equates electrostatic forces and London–van der Waals forces present at surfaces and is represented by the following equation

$$V_T\,(l) = V_A(l) + V_R(l)$$

where the total interaction energy (V_T) of a particle as a function of its separation distance (l) from a solid surface, is the sum of the van der Waals attraction (V_A) and the electrostatic interaction (V_R).[55] According to this theory, attraction of particles may occur when small distances of less than 1 nm between an approaching particle and a surface are evident or when a distance of 5 to 10 nm separates the particle in question and the surface.[56,57] These two regions are referred to as the primary minimum and the secondary minimum. Located between these two positions is an energy level where the surfaces experience maximum repulsion (an electrostatic repulsion occurs because the cell and the substratum surfaces both carry a negative charge). The magnitude of this is dependant upon the surface potential of the particle and the substratum, the separation distance, and the electrolitic strength of the aqueous medium. According to this theory, the net force of interaction arises from a balance between van der Waals forces of attraction and electrostatic double-layer forces (those which commonly have a repulsive effect). van der Waals attraction relates to the effective size of the bacterial cell which does not necessarily include the space occupied by appendages such as flagellum, pili, fimbriae, and exopolysaccharides. If these are present on the surface, they will serve to bridge the gap between the primary and secondary minimum, thereby increasing the effective distances over which forces will operate. Production of surface appendages is often subject to phase variation, with these appendages demonstrable in only a small fraction of actively growing culture. This may lead to situations where only a proportion of the population will immediately bind to a surface irreversibly, and where continued growth of the reversibly attached cells, expression of surface appendages, and exopolymer leads to a facilitated progression from the secondary to the primary minimum.[27] If this process is selected as the predictor of microbial adsorption, a number of problems may be encountered. These include the fact that this system was developed as a process applied to shear free systems which only exist within the boundary layer with most dynamic fluid systems experiencing a shear effect.[22] Also, geometrical considerations must be taken into account because, as mentioned previously, cellular appendages alter the cells' effective diameter near the surface and, hence, alter the repulsive effects experienced within the regions of maximal repulsion between the primary and secondary minimum.[58]

Busscher and Weerkamp[59] have offered a three-point hypothesis of bacterial adhesion which relates to the distance of the bacteria from the surface. At a distance of greater than 50 nm from the surface van der Waals forces exist. With a distance of 10 to 20 nm from the surface, van der Waals and electrostatic interactions occur, which are associated with reversible and irreversible adhesion. With a distance of less than 1.5 nm van der Waals, electrostatic and specific interactions occur between the bacteria and the surface, producing irreversible binding and the formation of exopolysaccharides.

The second system or theory which models the attachment of bacteria to a surface is based on the free energy system. The process suggests that if the total free energy of a system is reduced by cell contact with a surface, then adsorption of the cell to the substratum will occur.[60] More information about this process can be located elsewhere.[61]

The physico-chemical models of surface interaction assume that the surfaces are small, smooth, and energetically homogenous. This is a situation not true of bacteria.[62,63] Overall, these approaches fail to incorporate the microscopic condition of the cell's outer surface or adaptive microbial behaviour, preventing an explanation of all aspects of bacterial adhesion.[61]

To date, no satisfactory model is available to fully explain the adhesion process in turbulent flowing systems.

6.4.4 Extracellular Polymeric Substances (EPS) Involved in Biofilm Formation

If cells reside at a surface for a certain time, irreversible adhesion forms through the mediation of a cementing substance which is extracellular in origin. This extracellular material associated with the cell has been referred to as glycocalyx,[62] a slime layer, capsule, or sheath. Costerton et al.,[64] referred to the glycocalyx as, "those polysaccharide-containing structures of bacterial origin, lying outside the integral elements of the outer membrane of Gram-negative cells and peptidoglycan of Gram-positive cells."

The involvement of extracellular polymers in bacterial attachment has been documented for both fresh[65] and marine water bacteria.[27,66] Analysis of bacteria isolated from these environments has shown that the polymers produced are largely composed of acidic polysaccharides.[67] The extent to which the polysaccharides are involved in the adhesion process is, however, open to question. Some reports suggest roles of the polysaccharides both in the initial, reversible phase of adhesion[66,68] and the later, irreversible phase.[27,51,68] Some evidence has been presented suggesting that excess polymer production may even prevent adhesion, although trace amounts of polysaccharide might be required initially.[69] Although the association of exopolysaccharide with attached bacteria has been demonstrated by both electron microscopy[70,71] and light microscopy,[51,72] there is little evidence to suggest that extracellular polymeric substances (EPS) participate in the initial stages of adhesion, despite its synthesis by many species in the adherent population.

EPS seem to provide many benefits to a biofilm[73] including

1. Cohesive forces within the biofilm.
2. Absorbing nutrients, both organic and inorganic.[74,75]
3. Absorbing microbial products and other microbes.
4. Protecting immobilised cells from rapid environmental changes.
5. Absorbing heavy metals from the environment.
6. Absorbing particulate material.
7. Serving as a means of intercellular communication.
8. Enhancing intercellular transfer of genetic material.

Extracellular polymeric substances have also been shown to bind metal ions selectively[76] and to accelerate corrosion often owing to the lipopolysaccharides (LPS) present in the outer most layer of gram-negative bacterial cells. Research is still ongoing in this area, suggesting that this list is by no means exhausted.

Molecules other than polysaccharides and sugars have been found within the biofilm organic matrix. Examples include glycoproteins,[77] proteins, and nucleic acids. The polymers which constitute the biofilm are, however, dominated by polysaccharides with lesser amounts of proteins, nucleic acids, and others which are still in the process of being identified. Therefore, components of the organic matrix of the biofilm are generally referred to as EPS.[73]

The polysaccharides associated with EPS are known to help anchor the producing bacteria to the substratum by participation of their polyhydroxyl groups. Extending lengths of polymers attached to cell surfaces can interact with vacant bonding sites on the surface by polymer-bridging and, as a result, the cell is held near the surface. Possible mechanisms for polymer bridging have been suggested[73] but they are not fully understood. The bacterium through predominately covalent bonds connect it to the exopolymers, firmly attaching it to the substratum via exopolymer-substratum interactions. Interest in the ecology of sessile microbial populations has often focused on the extracellular polymers elaborated by the cells.[64,66,78] In aquatic habitats, microbial exopolymers commonly occur as discrete capsules firmly attached to the cell surface or as slime fibres loosely associated with or dissociated from the cells. While it is now believed that many of the capsular polymers may serve as holdfasts, anchoring cells to each other and to inert surfaces, the extent to which they facilitate other interactions between sessile bacteria and their environment is less understood.

A biofilm generally has a high content of EPS consisting of between 50 and 90% of the matrix.[73] An understanding of the physical and chemical characteristics of the biofilm matrix and its relationship to the organisms present is necessary for understanding of the structure and functioning of biofilms. EPS influence the physical properties of the biofilm, including diffusivity, thermal conductivity, and rheological properties. EPS, irrespective of charge density or its ionic state, have some of the properties of diffusion barriers, molecular sieves, and adsorbents, thus influencing physio-chemical processes such as diffusion and fluid frictional resistance. The predominantly polyanionic, highly hydrated nature of EPS also means that it can act as an ion exchange matrix, serving to increase local concentrations of ionic species such as heavy metals, ammonium, potassium, etc. while having the opposite effect on anionic groups. It has been reported to have no effect on uncharged potential nutrients, including sugars. However, bacteria are assumed to concentrate and use cationic nutrients such as amines, suggesting that EPS can serve as a nutrient trap, especially under oligiotrophic conditions.[64] Conversely, the penetration of charged molecules such as biocides and antibiotics may be, at least partly, restricted by this phenomenon.[79]

Other roles suggested for the biofilm extracellular matrix are as an energy store and site of both intracellular communication and genetic transfer.[73] The extracelluar matrix may contain particulate materials such as clays, organic debris, lysed cells, and precipitated minerals with the composition of different biofilms being dominated by different components. Biofilms, therefore, appear to vary dynamically with their extracellular matrix composition clearly changing with time.

6.4.5 Microcolony and Biofilm Formation

The adsorption of macromolecules and attachment of microbial cells to a substratum are only the first stages in the development of biofilms. This is followed by the growth of bacteria, development of microcolonies (Figure 6.4), recruitment of additional attaching bacteria, and often colonisation of other organisms, for example, microalgae. As attachment of bacteria takes place, the bacteria begin to grow and extracellular polymers are produced and accumulated so that the bacteria are eventually embedded in a hydrated polymeric matrix. The biofilm bacteria, consequently,

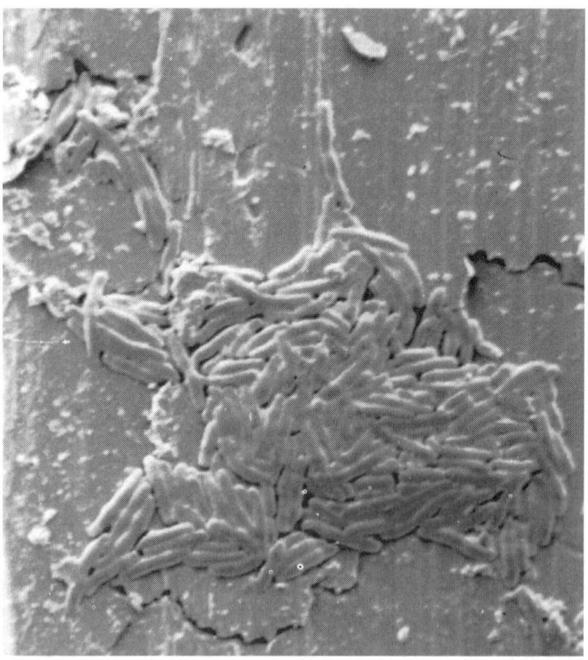

FIGURE 6.4 A microcolony on stainless steel present in potable water. Reprinted from *Water Research*, 32, Percival, S., Biofilms, mains water and stainless steel, 2187–2201, Copyright 1998, with permission from Elsevier Science.

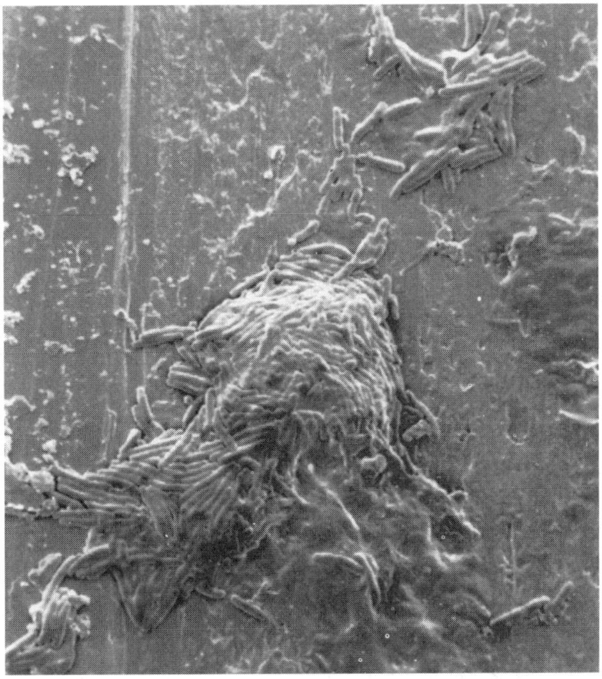

FIGURE 6.5 A microcolony of rod shaped bacteria-based encased in an amorphous gel of extracellular material. Reprinted from *Water Research,* 32, Percival, S., Biofilms, mains water and stainless steel, 2187–2201, Copyright 1998, with permission from Elsevier Science.

are immobilised and, thus, dependent upon substrate flux from the liquid phase and/or exchange of nutrients with their neighbours in the biofilm. An important feature of the biofilm environment is that the microorganisms are immobilised in relatively close proximity to one another (Figure 6.5).

Additional organisms may be located within or on top of the biofilm matrix. Specific functional types of organisms may, through their activities, create conditions that favour other complementary functional groups. This would lead to the establishment of spatially separated, but interactive, functional groups of bacteria, which exchange metabolites at group boundaries achieving physiological cooperation.[80] As biofilm communities tend to be complex both taxonomically and functionally, there is considerable potential for synergistic interaction among constituent organisms. There may be the development of homeostatic mechanisms that could protect the bacteria from outside perturbations. Such mechanisms for balance would be extremely important in natural communities exposed to disturbances such as pollution. As the biofilms' heterogeneity increases, chemical and physical microgradients develop which include pH, oxygen, and nutrient gradients.[81]

In biofilms located in natural environments, there is evidence of a high level of cellular interaction and competitive behavior.[82] This competition arises as a consequence of resource availability. It is well known that higher organisms will also influence the outcome of a maturing biofilm, particularly with the existence of grazing protozoa. As a result of competition strategies by specific species of bacteria, the biofilm system is under a constant flux.[28,48,81]

Microbial succession is a common feature of biofilms, particularly within natural systems. During adhesion, the pioneering or primary coloniser to any surface has defined requirements dictated by the conditioning film. The succession of the biofilm community is then governed by a number of physiological and biological events initiated by this pioneering species of bacteria.[28,68] Many researchers frequently have observed succession patterns of surface biofilms in both flowing and static systems.

It has been estimated that a mature biofilm contains only 10% or less of its dry weight in the form of cells.[83] Young biofilms generally contain few species, reflecting the low diversity of pioneering populations,[84] but this diversity increases to form a stable climax community and is often underestimated owing to the selectivity and inadequacy of pure-culture isolation techniques.[85]

As the biofilm develops, various gradients develop across it, as exchange of substances (nutrients and gasses) occurs on only one side.[73] A nutrient gradient develops, with aerobic respiration at the upper surface and fermentation in the middle layer with the resulting release of fermentation products such as ethanol, lactate, and succinate.[86] Generally, when the biofilm reaches a thickness of 10 to 25 μm, conditions at its base become anaerobic[87] indicating that the biofilm is now approaching a state of maturity, with a high species diversity and stability.[87] Under anaerobic conditions, anaerobic respiration may occur with, for example, sulphate reduction.

Surface characteristics are relevant during the buildup of a biofilm with surface roughness playing a significant role in the transport and adsorption of the first macromolecules and microbial cells to the surface.[16] Apart from increasing the available interfacial area, a rough surface enhances mass transfer coefficients and allows cells to anchor on its micro-irregularities where they are better protected from possible desorption. Regardless of surface roughness, the attachment of living particles is favourable energetically if the change in the free energy during the process is negative. In spite of metallic surfaces being favourable energetically to the attachment of the first cells, the chemical composition of surfaces may interfere with adhesion, cellular metabolism, and production of exopolymers.[88] The surface effect of certain metals on bacterial adhesion has been reported by Vieira et al.,[89] who found when counting the number of attached cells of *Pseudomonas fluorescens* on brass, copper, and aluminum surfaces after a few hours of exposure, aluminum surfaces were the most fouled, followed by copper and brass.

The structure of a biofilm within both mixed and pure culture systems evident in many different environments has been reported extensively in the literature. Intially, the biofilm was considered as an homogenous confluent structure being composed of a substratum, base film, and surface film exposed to a bathing fluid (Figure 6.6). However, research has now demonstrated that a biofilm exists as a heterogenous structure in a nonconfluent form.

With the use of confocal scanning laser microscopy together with microelectrode measurements, researchers have established that the biofilm consists of cell clusters which are discrete aggregates of cells located in an EPS matrix. These clusters have been shown to vary in shape, often ranging from cylinders to filaments and forming a mushroom structure.[90] Within these systems, owing to the evidence of water

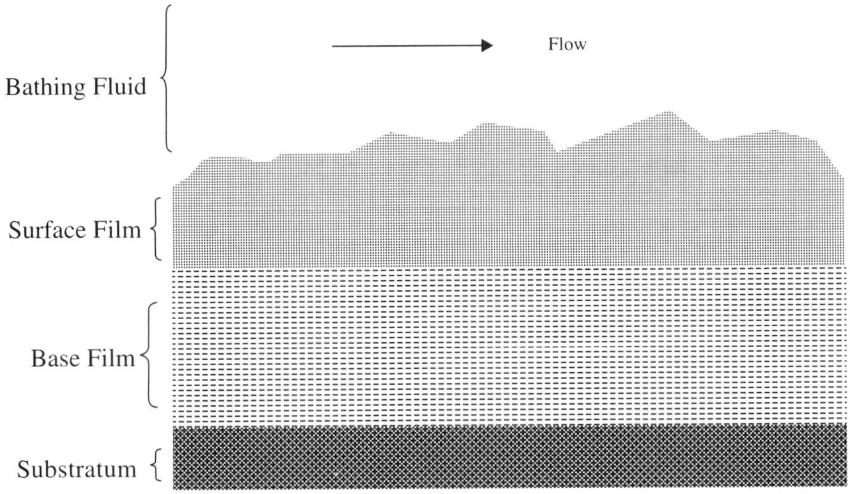

FIGURE 6.6 Homogeneous biofilm model.

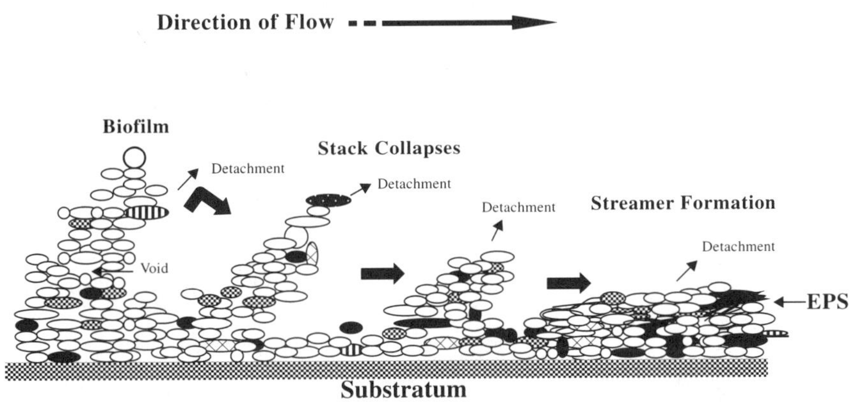

FIGURE 6.7 Diagram showing the development of streamers in potable water.

channels or patches of biofilms, a number of biofilm arrangements have been cited including aggregates, cell clusters, streamer, and stacks (Figure 6.7).[91-93] These open channels evident in biofilms are referred to as channels, voids, and pores. This would indicate that biofilms show a high degree of spatial and temporal complexity, particularly when present within potable water systems.

Therefore, the present conceptual model of a biofilm is described as cell clusters or stacks[94] separated by interstitial voids.[90] The evidence of voids facilitates mass transfer which favours higher concentrations of nutrients in the void spaces and also allows for cellular metabolites and by-products to be more concentrated under cell clusters. These stack systems, which are evident within oligiotrophic environments, have been replicated in simple computer simulations.[95]

Overall, the development of a biofilm is generally governed by a number of parameters[96] and include

1. Ambient and system temperatures which are related to season, day length, climate, and wind velocity.
2. Hydrodynamic conditions (shear forces, friction drag, and mass transfer).
3. Nutrient availability (concentration, reactivity, antimicrobial properties).
4. Roughness, hydrophobicity, and electrochemical characteristics of the surface.
5. pH (an approximately neutral pH of the water is optimal for the growth of most biofilm-forming bacteria).
6. The presence of particulate matter (this can become entrapped in the developing biofilm and provide additional attachment sites).
7. Effectiveness of biofilm control measures.

From the preceding list, we can see that many parameters play a role in affecting and also determining the structure of a biofilm. Overall, there are generally four major factors which influence biofilm structure.[4] These include the surface or interface properties, hydrodynamics, nutrients, and biofilm consortia. This list is by no means exhaustive but reflects the large numbers of factors that affect the developing biofilm. The controversial condition known to affect biofilm structure includes the hydrodynamic forces known to operate within flowing conditions. It is now well established that biofilms exposed to high turbulent flow experience and develop a phenomena known as streaming (Figure 6.8). The significance of this is still under study.

6.4.6 Detachment from the Biofilm

Overall, detachment can be perceived as consisting of five processes according to Bryers.[97] These include

1. Erosion (single cells).
2. Sloughing (clusters of cells).
3. Abrasion.
4. Human intervention.
5. Predator grazing.

Erosion, sloughing, and abrasion are defined as physical processes. In general, erosion is classified as the removal of small particles of biofilms as a result of shear forces generated by fast flowing fluids. Some research, however, suggests that detachment is independent of shear stress but dependent on mass transfer of the nutrients to the biofilm. This suggests that there will be a flow velocity region where detachment rate increases with increasing flow velocity. Generally though, it is found that the shear effect of water causes the continuous removal of small sections of biofilms. As a rule, erosion increases with increasing biofilm thickness. On newly formed immature biofilms, this type of detachment is not often evident.[98]

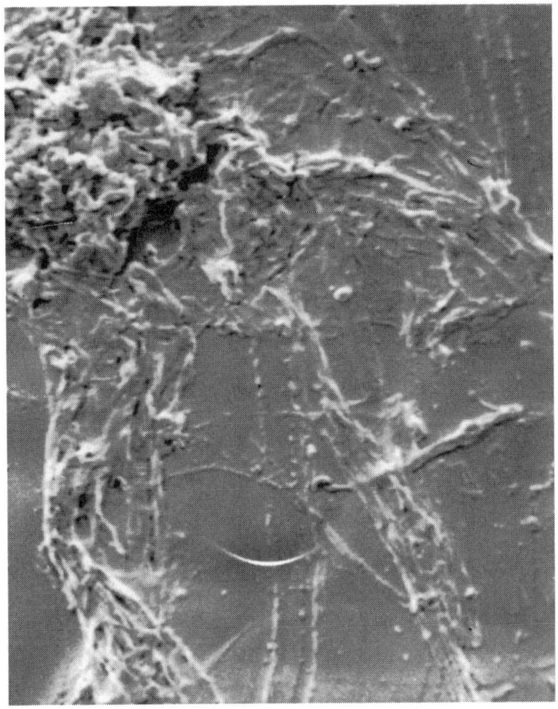

FIGURE 6.8 A streamer evident in potable water. Reprinted from *Water Research,* 32, Percival, S., Biofilms, mains water and stainless steel, 2187–2201, Copyright 1998, with permission from Elsevier Science.

In contrast, sloughing is referred to as being a random and discrete process,[49] involving the detachment of large particles of biofilm frequently evident within thicker biofilms, particularly within nutrient rich environments.[73] Sloughing often occurs in older and thicker biofilms and involves random massive removal of biofilm usually owing to nutrient or oxygen depletion within the biofilm or some dramatic change in the immediate environment.[49,99] It is also possible that sloughing might be physiologically mediated through the activation or induction of certain enzymes.[100]

Abrasion, on the other hand, is caused by thc collision of solid particles with the biofilm and human intervention involving detachment of the biofilm by chemical or physical means.[52] Finally, predator grazing is the consumption of biofilms by organisms such as protozoa, snails, and worms known to be particularly evident in fresh and marine water environments.

In 1990, Characklis et al.[22] recategorised the process of detachment into only three areas, namely erosion, sloughing, and abrasion. They referred to detachment as an interfacial transfer process which involved the transfer of cells and other components from the biofilm compartment to the bulk liquid with the detachment of microbial cells and related biofilm material occurring from the moment of initial attachment.

Other factors known to affect detachment are environmental parameters including pH, temperature, and the presence of organic macromolecules either absorbed on the substratum or dissolved in the liquid phase.[101] The effects these conditions have on bacterial detachment are generally species specific.

Overall, the accumulation of a biofilm is the net result of processes that produce biomass and processes that remove it. The accumulation continues until the biofilm reaches a steady state where the product of the biomass is equal to biomass detachment. The overall net accumulation of a biofilm associated with a surface can be determined by the following equation developed by Trulear and Characklis[102]

$$\text{net rate attached biofilm accumulation} = \text{rate of biomass production} - \text{rate of biomass detachment}$$

Surface roughness of the substratum may also be a significant factor in biofilm detachment, with early events in biofilm formation being controlled by hydrodynamic forces.[103] As detachment increases with increasing fluid shear stress at the substratum surface, macro- and microroughness may significantly influence detachment rates of the biofilm owing to a sheltering effect from hydrodynamic shear. The detached cells may be transported close to the surface (in the viscous sublayer) resulting in collisions with the surface and providing more opportunity for reattachment.

To date, detachment is a poorly understood phenomena which complicates the formation of satisfactory models. There is poor correlation between detachment and shear force[17,104] and between shear and biofilm thickness, with the growth rate of biofilm influencing the ease with which a biofilm detaches. Previously, it was considered that turbulent bursts transcending the viscous sublayer were responsible for generating forces necessary to remove a biofilm from a surface. Now, it is thought that such bursts do not have sufficient power to achieve this; research indicates that the biofilms are viscoelastic, not rigid, which seems to provide resistance to turbulent bursts.[105]

Despite lack of research in this area, detachment of biofilms from surfaces into surrounding environments does have very important implications within the manufacturing, medical, and public arenas. Whilst the phenomena of biofilm detachment does have implications on biofilm development and survival, it also has implications in relation to infection, contamination, and public health issues particularly in potable water supplies. In microbiological terms, detachment from surfaces may seem at first to be a disadvantage in biofilm development. However, detachment has important implications in biofilm formation. It is found that biofilms with greater detachment rates have been found to have larger fractions of active bacteria. It has also been reported that detachment can occur as a result of low nutrient conditions, indicating some survival mechanism which may be genetically determined. Therefore, detachment is not just important for promoting genetic diversity but also for escaping unfavorable habitats aiding in the development of new niches.

However, in relation to the public's health, the detachment process has profound implications upon waterborne diseases, aetiology, factory hygiene, and, ultimately, the quality of products which may contain a higher than normal microbial loading supplied commercially to the disconcerting consumer.

6.5 REFERENCES

1. Characklis, W. G. and Marshall, K. C., 1990, *Biofilms,* John Wiley & Sons, New York.
2. Hamilton, W. A., 1985, Biofilms and microbially influenced corrosion, in *Microbial Biofilms,* Lappin-Scott, H. M. and Costerton, J. W., Eds., Cambridge University Press, Cambridge, 171.
3. Costerton, J. W., Lewandowski, Z., Caldwell, D. E., Korber, D. R., and Lappin-Scott, H. M., 1995, Microbial biofilms, *Ann. Rev. Microbiol.*, 49, 711.
4. Stoodley, P., Boyle, J. D., Dodds, I., and Lappin-Scott, H. M., 1997, Consensus model of biofilm structure, in *Biofilms: Community Interactions and Control,* Third Meeting of the British Biofilm Club, Gregynog Hall, Powys, September 26–28, 1997, 1.
5. Goodman, A. E., and Marshall, K. C., 1995, Genetic responses of bacteria at surfaces, in *Microbial Biofilms,* Lappin-Scott, H. M. and Costerton, J. W., Eds., Cambridge University Press, Cambridge, 80.
6. McCoy, W. F. and Costerton, J. W., 1982, Fouling biofilm development in tubular flow systems, *Dev. Ind. Microbiol.*, 23, 551.
7. Fletcher, M., 1991, The physiological activity of bacteria attached to solid surfaces, *Adv. Microb. Physiol.*, 32, 53.
8. Costerton, J. W., Cheng, K. J., Geesey, G. G., Ladd, T. I. M., Nickel, J. C., Dasgupta, M., and Marie, T. J., 1987, Bacterial biofilms in nature and disease, *Ann. Rev. Microbiol.,* 41, 435.
9. Hermansson, M. and Marshall, K. C., 1985, Utilization of surface localised substrate by non-adhesive marine bacteria, *Microb. Ecol.*, 11, 91.
10. Costerton, J. W. and Lappin-Scott, H. M., 1989, Behaviour of bacterial biofilms, *Am. Soc. Microbiol. News,* 55, 650.
11. Anwar, H., Dasgupta, M., Lam, K., and Costerton, J. W., 1992, Establishment of aging biofilms: possible mechanism of bacterial resistance to antimicrobial therapy, *Antimicrob. Agents Chemother.,* 36, 1347.
12. van Loosdrecht, M. C. M., Lyklema, J., Norde, W., and Zehnder, A. J. W., 1990, Influence of interfaces on microbial activity, *Microbiol. Rev.,* 54, 75.
13. Block, J. C., Haudidier, K., Paquin, J. L., Miazga, J., and Levi, Y., 1993, Biofilm accumulation in drinking water distribution systems, *Biofouling,* 6, 333.
14. Kjelleberg, S., Humphrey, B. A., Marshall, K. C., and Jones, G. W., 1983, Initial phases of starvation and activity of bacteria at surfaces, *Appl. Environ. Microbiol.,* 46, 978.
15. Decho, A. W., 1990, Microbial exopolymer secretions in ocean environments: their role(s) in food webs and marine processes, *Oceanogr. Mar. Biol. Ann. Rev.,* 28, 73.
16. Percival, S. L., Knapp, J. S., Edyvean, R., and Wales, D. S., 1997, Biofilm development on 304 and 316 stainless steels in a potable water system, *J. Inst. Water Environ. Manage.,* 11, 289.
17. Percival, S. L., Knapp, J. S., Wales, D. S., and Edyvean, R., 1998, The effects of the physical nature of stainless steel grades 304 and 316 on bacterial fouling, *Br. Corros. J.*, 33, 121.
18. van Loosdrecht, M. C. C., Lyklema, J., Norde, W., and Zehnder, A. J. W., 1989, Bacterial adhesion: a physico-chemical approach, *Microb. Ecol.,* 17, 1.
19. Marshall, K. C., 1976, *Interfaces in Microbial Ecology,* Harvard University Press, Cambridge, MA.
20. Marshall, K. C., 1985, Mechanisms of bacterial adhesion at solid-water interfaces, in *Bacterial Adhesion,* Savage, D. C. and Fletcher, M., Eds., Plenum Press, New York, 133.

21. Bryers, J. D., 1987, Biologically active surfaces: processes governing the formation and persistence of biofilms, *Biotechnology,* 3, 57.
22. Characklis, W. G., McFeters, G. A., and Marshall, K. C., 1990, Physiological ecology of biofilm systems, in *Biofilms*, Characklis, W. G. and Marshall, K. C., John Wiley & Sons, New York, 341.
23. Loeb, G. I. and Neihof, R. A., 1975, Marine conditioning films, *Adv. Chem. Ser.*, 145, 319.
24. Neihof, R. A. and Loeb, G. I., 1972, The surface charge of particulate matter in seawater, *Limnol. Oceanogr.*, 17, 7.
25. Neihof, R. and Loeb, G., 1974, Dissolved organic matter in seawater and the electric charge of immersed surfaces, *J. Mar. Res.*, 32, 5.
26. Chamberlain, A. H. L., 1992, The role of adsorbed layers in bacterial adhesion, in *Biofilms — Science and Technology,* Melo, L. F., Bott, T. R., Fletcher, M., and Capdeville, B., Eds., Alvor, Portugal, May 18–29, Kluwer Academic Publishers, London, 59.
27. Marshall, K. C., Stout, R., and Mitchell, R., 1971, Mechanism of the initial events in the sorption of marine bacteria to surfaces, *J. Gen. Microbiol.,* 68, 337.
28. Baier, R. E., 1984, Initial events in microbial film formation, in *Marine Biodetermination: An Interdisciplinary Approach,* Costlow, J. D. and Tipper, R. C., Eds., E & F. N. Spon, London, 57.
29. Rittle, K. H., Helmstetter, C. E., Meyer, A. E., and Baier, R. E., 1990, *Escherichia coli* retention on solid surfaces as functions of substratum surface energy and cell growth phase, *Biofouling*, 2, 121.
30. Paerl, H. W., 1975, Microbial attachment to particles in marine and freshwater ecosystems, *Microb. Ecol.,* 2, 73.
31. Dazzo, F. B., Truchet, G. L., Sherwood, F. E., Hrabak, E. M., Abe, M., and Pankratz, S. H., 1984, Specific phases of root hair attachment in the *Rhizobium trifolii*-clover symbiosis, *Appl. Environ. Microbiol.,* 48, 1140.
32. Vesper, S. J. and Baer, W. D., 1986, Role of pili (fimbria) in attachment of *Bradyrhizobium japonicum* to soyabean roots, *Appl. Environ. Microbiol.,* 52, 134.
33. Sjollema, J., van der Mei, H. C., Uyen, H. M. W., and Busscher, H. J., 1990, The influence of collector and bacterial cell surface properties on the deposition of oral *Streptococci* in a parallel: plate flow cell, *J. Adh. Sci. Technol.,* 4, 765.
34. Characklis, W. G., 1981, Fouling biofilm development: A process analysis, *Biotechnol. Bioeng.,* 23, 1923.
35. Munson, B. R., Young, D. F., and Okishi, T. H., 1990, Fundamental fluid mechanics, in *Fundamental Fluid Mechanics,* John Wiley & Sons, London.
36. Fletcher, M. and Marshall, K. C., 1982, Are solid surfaces of ecological significance to aquatic bacteria?, *Adv. Microb. Ecol.,* 12, 199.
37. Lappin-Scott, H. M., Jass, J., and Costerton, J. W., 1993, *Microbial Biofilm Formation and Characterisation,* Society for Applied Bacteriology Technical Series No. 30, Blackwell Science.
38. Calwell, D. E. and Lawrence, J. R., 1988, Study of attached cells to continous-flow slide culture, in *A Handbook of a Laboratory Model System for Microbial Ecosystem Research,* Wimpenny, W. T., Eds., CRC Press, Boca Raton, 117.
39. Brading, M. G., Jass, J., and Lappin-Scott, H. M., 1995, Dynamics of bacterial biofilm formation, in *Microbial Biofilms,* Lappin-Scott, H. M. and Costerton, J. W., Eds., Cambridge University Press, London.
40. Massey, B. S., 1989, Mechanisms of fluids, 6th ed., Chapman & Hall, London, 148.
41. Caldwell, D. E., Korber, D. R., and Lawrence, J. R., 1992, Confocal laser microscopy and digital image analysis in microbial ecology, *Adv. Microb. Ecol.,* 12, 1.

42. Cleaver, J. W. and Yates, B., 1975, A sublayer model for the deposition of particles from a turbulent flow, *Chem. Eng. Sci.,* 30, 983.
43. Speilman, L. A., 1977, Particle capture from low speed laminar flows, *Ann. Rev. Fluid Mech.,* 9, 297.
44. Young, L. Y. and Mitchell, R., 1973, The role of chemotactic responses in primary microbial film formation, in *Proceedings of the 3rd International Congress on Marine Corrosion and Fouling,* NACE, Houston, TX, 617.
45. Young, L. Y. and Mitchell, R., 1973, Negative chemotaxis of marine bacteria to toxic chemicals, *Appl. Environ. Microbiol.,* 34, 434.
46. Fletcher, M., 1977, The effects of culture concentration and age, time, and temperature on bacterial attachment to polystyrene, *Can. J. Microbiol.,* 23, 1.
47. Marmur, A. and Ruckenstein, E., 1986, Gravity and cell adhesion, *J. Colloid Interface Sci.,* 114, 261.
48. Wahl, M., 1989, Marine epibiosis. 1. Fouling and antifouling: some basic aspects, *Mar. Ecol. Prog. Ser.,* 58, 175.
49. Applegate, D. H. and Bryers, J. D., 1991, Effects of carbon and oxygen limitations and calcium concentrations on biofilm removal processes, *Biotechnol. Bioeng.* 37, 17.
50. Watkins, L. and Costerton, J. W., 1984, Growth and biocide resistance of bacterial biofilms in industrial systems, *Chem. Times Trends,* October, 35.
51. Zobell, C. E., 1943, The effect of solid surfaces upon bacterial activity, *J. Bacteriol.,* 46, 39.
52. Rittman, B. E., 1989, The effect of shear stress on biofilm loss rate, *Biotechnol. Bioeng.,* 24, 501.
53. Derjaguin, B. V. and Landau, L., 1941, Theory of the stability of strongly charged lyophobic sols and of adhesion of strongly charged particles in solution of electrolytes, *Acta Physiochim. URSS,* 14, 633.
54. Verwey, E. J. W. and Overbeek, J. T. G., 1948, *Theory of the Stability of Lyophobic Colloids,* Elsevier, Amsterdam.
55. van Loosdrecht, M. C. M., Norde, W., and Zehnder, A. J. B., 1987, Influence of cell surface characteristics on bacterial adhesion to solid surfaces, in *Proceedings of the 4th European Congress on Biotechnology,* European Federation for Biotechnology, Brussels, Belgium, 575.
56. Bowen, B. D. and Epstein, N., 1979, Fine particle deposition in smooth parallel-plate channels, *J. Colloid Interface Sci.,* 72, 81.
57. Characklis, W. G., Turakhia, M. H., and Zelver, N., 1990, Transfer and interfacial transport phenomena, in *Biofilms,* Characklis, W. G. and Marshall, K. C., Eds., John Wiley & Sons, New York, 265.
58. Sjollema, J., van der Mei, H. C., Uyen, H. M., and Busscher, H. J., 1990, Direct observations of cooperative effects in oral streptococcal adhesion to glass by analysis of the spatial arrangement of adhering bacteria, *FEMS Microbiol. Lett.,* 69, 263.
59. Busscher, H. J. and Weerkamp, A., 1987, Specific and non-specific interactions: role in bacterial adhesion to solid substrata, *FEMS Microbiol. Rev.,* 46, 165.
60. Absolom, D. R., Lamberti, F. V., Policova, Z., Zingg, W., van Oss, C. J., and Neumann, A. W., 1983, Surface thermodynamics of bacterial adhesion, *Appl. Environ. Microbiol.,* 46, 90.
61. Korber, D. R., Lawrence, J. R., Lappin-Scott, H. M., and Costerton, J. W., 1995, Growth of microorganisms on surfaces, in *Microbial Biofilms*, Lappin-Scott, H. M. and Costerton, J. W., Eds., Cambridge University Press, London.
62. Costerton, J. W., Geesey, G. G., and Cheng, K. J., 1978, How bacteria stick?, *Sci. Am.,* 238, 86.

63. Busscher, H. J., Bellon-Fontaine, M. N., Sjollema, J., and van Der Mei, H. C., 1990, Relative importance of surface free energy as a measure of hydrophobicity in bacterial adhesion to solid surfaces, in *Microbial Cell Surface Hydrophobicity*, Doyle, R. J. and Rosenberg, M., Eds., 335.
64. Costerton, J. W., Irvin, R. T., and Cheng, K. J., 1981, The bacterial glycocalyx in nature and disease, *Ann. Rev. Microbiol.,* 35, 299.
65. Jones, C. H., Roth, I. L., and Sanders, W. M., 1969, Electron microscope study of a slime layer, *J. Bacteriol.,* 99, 316.
66. Corpe, W. A., 1970, An acid polysaccharide produced by a primary film forming marine bacterium, *Dev. Ind. Microbiol.,* 11, 402.
67. Fletcher, M., 1980, The question of passive versus active attachment mechanisms in non-specific bacterial adhesion, in *Microbial Adhesion to Surfaces to Surfaces*, Berkeley, R. C. W., Ed., Horwood Limited, Chichester, 67.
68. Fletcher, M. and Loeb, G. I., 1979, The influence of substratum characteristics on the attachment of a marine *Pseudomonas* to solid surfaces, *Appl. Environ. Microbiol.,* 37, 67.
69. Brown, C. M., Ellwood, D. C., and Hunter, J. R., 1977, Growth of bacteria at surfaces: influence of nutrient limitations, *FEMS Microbiol. Lett.,* 1, 163.
70. Geesey, G. G., Richardson, W. T., Yeomans, H. G., Irvin, R. T., and Costerton, J. W., 1977, Microscopic examination of natural sessile bacterial populations from an alpine stream, *Can. J. Microbiol.,* 23, 1733.
71. Dempsey, M. J., 1981, Marine bacterial fouling: a scanning electron microscope study, *Mar. Biol.,* 61, 305.
72. Allison, D. G. and Sutherland, I. W., 1984, A staining technique for attached bacteria and its correlation to extracellular carbohydrate production, *J. Microbiol. Methods,* 2, 93.
73. Characklis, W. G. and Cooksey, K. E., 1983, Biofilms and microbial fouling, *Adv. Appl. Microbiol.,* 29, 93.
74. Bryers, J. D., 1984, Biofilm formation and chemostat dynamics: pure and mixed culture considerations, *Biotechnol. Bioeng.,* 26, 948.
75. Marshall, K. C., 1992, Biofilms: an overview of bacterial adhesion, activity and control at surfaces, *Am. Soc. Microbiol. News,* 58, 202.
76. Ford, T. E., Maki, J. S., and Mitchell, R., 1988, Involvement of bacterial exopolymers, in *Metal Ions and Bacteria,* Beveridge, T. J. and Doyle, R. J., Eds., Wiley-Interscience, New York, 257.
77. Humphrey, B. A., Dickson, M. R., and Marshall, K. C., 1979, Physiological and *in situ* observations on adhesion of gliding bacteria to surfaces, *Arch. Microbiol.,* 120, 231.
78. Uhlinger, D. J. and White, D. C., 1983, Relationship between physiological status and formation of extracellular polysacharide glycocalyx in *Pseudomonas atlantica, Appl. Environ. Microbiol.,* 45, 64.
79. Costerton, J. W. and Lashen, E. S., 1984, The influence of biofilm efficacy of biocides on corrosion-causing bacteria, *Mat. Perform.,* 23, 34.
80. Blenkinsopp, S. A. and Costerton, J. W., 1991, Understanding bacterial biofilms, *Trends Biotechnol.,* 9, 138.
81. Connell, J. H. and Slatyer, R. O., 1977, Mechanisms of succession in natural communities and their role in community stability and organization, *Am. Nat.,* 111, 1119.
82. Fredrickson, A. G., 1977, Behaviour of mixed cultures of microorganisms, *Ann. Rev. Microbiol.,* 33, 63.

83. Hamilton, W. A., 1985, Sulphate-reducing bacteria and anaerobic corrosion, *Ann. Rev. Microbiol.,* 39, 195.
84. Atlas, R. M., 1984, Diversity of microbial communities, *Adv. Microb. Ecol.,* 7, 1.
85. Brozel, V. S. and Cloete, T. E., 1990, Evaluation of agar plating methods for the enumeration of viable aerobic heterotrophs in cooling water, *6th Biennial Congress of the South African Society for Microbiology,* the South African Society Abstracts 22.13.
86. Pfennig, N., 1984, Microbial behaviour in natural environments, in *The Microbe,* Part 11, *Prokaryotes and Eukaryotes,* Oxford University Press, Oxford.
87. Hamilton, W. A., 1987, Biofilms: microbial interactions and metabolic activities, in *Ecology of Microbial Communities,* Fletcher, M., Gray, T. R. G., and Jones, J. G., Eds., Oxford University Press, Oxford, 361.
88. Beech, I. B. and Gaylarde, C. C., 1992, Attachment of *Pseudomonas fluorescens* and *Desulfovibrio* to mild steel and stainless steel—first step in biofilm formation, Sequeira, A. C. and Tiller, A. K., Eds., *Proceedings of the 2nd European Federation of Corrosion,* Portugal, 1991, European Federation of Corrosion Publication No. 8, The Institute of Materials, Portugal, 61.
89. Vieira, M. J., Oliveira, R., Melo, L., Pinheiro, M., and van der Mei, H., 1992, Adhesion of *Pseudomonas fluorescens* to metallic surfaces, *J. Dispersive Sci. Technol.,* 13(4), 437.
90. Lewandowski, Z., Stoodley, P., and Roe, F., 1995, Internal mass transport in heterogeneous biofilms. Recent advances, in Corrosion/95, Paper No. 222, NACE International, Houston, TX.
91. Costerton, J. W., Lewandowski, Z., de Beer, D., Calwell, D., Korber, D. R., and James, G., 1994, Biofilms, the customised microniches, *J. Bacteriol.,* 176, 2137.
92. DeBeer, D., Stoodley, P., Roe, F., and Lewandowski, Z., 1994, Effects of biofilm structures on oxygen distribution and mass transfer, *Biotechnol Bioeng.,* 43, 1131.
93. Gjaltema, A., Arts, P. A. M., van Loosdrecht, M. C. M., Kuenen, J. G., and Heijinen, J. J., 1994, Heterogeneity of biofilms in rotating annual reactors: occurrence, structure and consequences, *Biotechnol. Bioeng.,* 44, 194.
94. Geesey, G. G., Characklis, W. G., and Costerton, J. W., 1992, Centers, new technologies focus on biofilm heterogeneity, *ASM News,* 58(10), 546.
95. Wimpenny, J. W. T. and Colasanti, R., 1997, A unifying hypothesis for the structure of microbial biofilms based on cellular automaton models, *FEMS Microbiol. Ecol.,* 22, 1.
96. Wolfaardt, G. M., Archibald, R. E. M., and Cloete, T. E., 1990, Techniques for biofouling monitoring during alkaline paper manufacture, TATTSA 90, *Conference Proceedings,* Technical Association of the Pulp and Paper Industry, South Africa.
97. Bryers, J. D., 1987, Biologically active surfaces: processes governing the formation and persistence of biofilms, *Biotechnol. Prog.,* 3, 57.
98. Chang, H. T. and Rittman, B. E., 1988, Comparative study of biofilm shear loss on different adsorptive media, *J. Water Pollut. Control Fed.,* 60, 362.
99. Howell, J. A. and Atkinson, B., 1976, Sloughing of microbial film in trickling filters, *Water Res.,* 10, 307.
100. Boyd, A. and Chakrabarty, A. M., 1994, Role of alginate lyase in cell detachment of *Pseudomonas aeruginosa, Appl. Environ. Microbiol.,* 60, 2355.
101. McEldowney, S. and Fletcher, M., 1988, Effect of pH, temperature, and growth conditions on the adhesion of a gliding bacterium and three nongliding bacteria to polystyrene, *Microb. Ecol.,* 16, 183.

102. Trulear, M. G. and Characklis, W. G., 1982, Dynamics of biofilm processes, *J. Water Pollut. Control Fed.*, 54, 1288.
103. Powell, M. S., and Slater, N. K. H., 1982, Removal rate of bacterial cells from glass surfaces by fluid shear, *Biotechnol. Bioeng.*, 24, 2527.
104. Cooksey, K. E., 1992, Extracellular polymers in biofilms, in *Biofilms-Science and Technology*, Melo, L. F., Bott, T. R., Fletcher, M., and Capdeville, B., Eds., Kluwer Academic Publishers, London, 137.
105. Characklis, W. G., 1980, Biofilm development and destruction, Final Report EPRI Cs-1554, Project RP 902-1, Electric Power Research Institute, Palo Alto, CA.

7 Biofilm Formation in Potable Water

CONTENTS

7.1 INTRODUCTION

The presence of biofilms in potable water has been evident since 1930 with the reported problem of the existence of the regrowth of *Bacterium coli*.[1,2] Another 7 years passed until research confirmed that microbes were capable of multiplication on the walls of water distribution pipes.[3] However, it took another 42 years for the presence of biofilms associated with potable water to generate concerns when associated with pipe tubercles and encrustations[4] indicating some form of public health problem. Even today, the presence of biofilms within potable water pipes are being observed by a number of researchers using more modern methods, particularly the scanning electron microscope.[5-7] Despite accumulating evidence confirming the presence of biofilms in potable water, it is only recently that any health implications associated with their growth are being documented, particularly as they are known to provide a safe haven for the growth of pathogens, opportunistic pathogens, viruses, and protozoa.[8]

The study of biofilm development in potable water is of great interest to drinking water companies in relation to the effects they have on both the aesthetics and the public's health. In terms of the aesthetics, it is widely acknowledged that potable water can be affected by organisms such as fungi and *Actinomycetes* which are well known to cause taste and odour problems.[9,10] These have been documented as being found as part of biofilms which may well exasperate the taste and odour problems experienced in potable water. It has also been shown that some bacteria found as part of a biofilm can lead to corrosion of pipe material.[11] However, by far the most important concern to a water company is the fact that biofilms can and do have the potential of leading to both heterotrophic and coliform regrowth and providing a safe haven for pathogens and opportunistic pathogens.[12] This may well have very important implications in factorising the increasing numbers of gastroenteritis and noscomial infections being both reported and unreported worldwide.

The adhesion of bacteria to surfaces in potable water have been shown to be a major determinant in protecting microorganisms from chlorine. Therefore, sloughing of the biofilm may well serve as a system for the dissemination of potentially problematic microbes. Despite this possibility, biofilms evident on the pipe walls of potable water systems are not examined routinely as part of an assessment procedure in potable water systems. This is possibly owing to the fact that long runs of water distribution lines cannot be sampled because of the inaccesibility of the water distribution systems. Research that is, therefore, carried out and documented on biofilms has generally used laboratory simulations and small pilot rig systems which cannot truly mimic a water distribution system faultlessly, owing to the presence of so many changing variables which present themselves in these systems. However, it is now accepted that all materials which are exposed to a bathing fluid will support a biofilm and growth of that biofilm. In fact, to date no material has yet been developed which does not allow the adhesion of bacteria.

7.2 ADHESION AND BIOFILMS IN POTABLE WATER

As a material comes into contact with potable water, organic and inorganic particles from this bathing fluid become adsorbed onto pipe surfaces and are then followed by microorganisms. Once microbes are attached to a conditioned surface, aided by the use of EPS, fimbria, and flagella, the microbes grow and multiply, consuming the nutrients evident within the conditioning film. As the water flows past the biofilm, nutrients can diffuse into the biofilm by simple diffusion, allowing for biofilm maturation.

Once bacteria are attached to a surface they may behave in two ways[13]

1. Bacteria are not able to multiply in the potable water system and thus are washed out from the network system (water + biofilm) governed by the flow rate and the dilution rate.
2. Bacteria are able to multiply in the potable water system and remain depending on availability of nutrients. Research has found that eliminating suspended cells entering the network will decrease the biofilm population.[14]

Nutrients are a necessity for the growth of bacteria. These nutrients are classed as biodegradable, measured as either biodegradable organic carbon (BDOC)[15] or assimible organic carbon (AOC). A number of models are available to predict both the impact of BDOC on the biological stability of the waters and the numbers of bacteria in the planktonic and sessile states.[14,16,17] The consequence of the BDOC is correlated with an increase in the number of cells in the biofilm and, subsequently, in the water phase.

7.3 MICROBIOLOGY OF POTABLE WATER AND BIOFILMS

Potable water generally contains a lot of microorganisms, particularly ones which are injured and dormant, as a result of the disinfection process (chlorination) and water treatment processes. However, this is a poorly defined area, where either the nature of the injury or the degree of dormancy are unknown. The only certain fact is that these injured or dormant bacteria are metabolically active but incapable of replication on culture media, with available culture methods for waterborne bacteria giving results well below the actual total cell count and which vary, however slightly, with the culture method utilised. It is widely acknowledged that the use of agar media for estimating bacteria in water samples produces results which generally underestimate bacteria in water samples.[18] This has led to the development of new rapid techniques for the identification of sessile bacteria.[19]

The majority of bacteria typical of freshwater ecosystems can be found among the flora of water distribution systems, although such organisms are present at very low concentrations. The water flora can be quantified by counting, using an epifluorescence microscope, or by culturing on nutrient agars (total heterotrophic plate counts). Very often, potable water is dominated by gram-negative bacteria,[6,20,21] particularly *Pseudomonas* and yellow/brown pigmented organisms of *Flavobacterium*-like species.[6,22-24] As well as regular harmless flora, pathogenic or potentially pathogenic organisms may be detected (e.g., *Legionella*) which could have serious consequences.

At least 50% of bacteria carried by the water are present as aggregates of dimensions greater than 5 μm or attached to nonbiological particles of over 5 μm in diameter.[5,25] These particles, as soon as they come into contact with water, become colonised by microorganisms from the aquatic environment. This biomass (bacteria, microscopic fungi, protozoa, and yeasts) is attached to either the walls of the pipework in a heterogeneous pattern determined by the shear forces acting on the wall of the precipitates, deposits of sediment, encrustations, or tubercles which form on the pipe wall surface.[4,5]

Although biofilms evident in potable water systems are patchy, a mature biofilm can be up to 200 μm thick which may cause a drop in oxygen levels resulting in anoxic conditions below the surface layer. Such conditions could lead to the growth of sulphate reducing bacteria (SRB) whose activity may corrode and pit the insides of pipelines.[26]

Other effects as a result of biofilms in potable water systems include

- Positive bacteriological tests (coliforms).
- High plate counts.
- Growth of opportunistic pathogens.
- Increased chlorine demands.

- Discolouration of water.
- Taste and odour effects.
- Growth of invertebrates.
- Corrosion and pitting.

Control of the biofilm becomes one of the main objectives of good potable water distribution practice. However, as chlorination does not prevent accumulation of biofilms[27,28] other preventive measures such as mechanical cleaning should be combined with the use of biocides for a greater biofilm removal rate and thus a longer lag phase to regrowth. It should also be noted that chlorination may also induce biofilm development in potable water.[29]

The characteristic of a biofilm, that is, species present, number of cells, specific gravity, and thickness, are controlled by a myriad of factors including number and diversity of the species present in the water, the concentration and nature of the biodegradable organic matter in the water, the hydraulic regime to which the system is subjected, and the characteristics of the support material colonised by the bacteria.[13]

Observations of biofilms in water distribution systems indicate little qualitative or quantitative uniformity. Although some bacterial species are more frequently detected than others, a fairly large variation in population composition can be seen. Cell numbers and distribution in pipe wall linings and corrosion products vary widely. Considering large differences in the water distribution system environments, it is not surprising that differences exist in the quality of the treatment plant effluent and in the structure and chemical composition of the pipe wall.

Within potable water biofilms a large diversity of microorganisms are generally found, most of which may be potentially harmless with possible exceptions (see Chapter 8). The presence of these bacteria suggests that the disinfection process is not inactivating the microorganisms and, therefore, will allow the continual development of biofilms within potable water pipes. This may result in deterioration in water quality hindering the detection of indicator organisms often leading to public health implications. Research suggests that biofilm formation is possibly induced as a result of the low nutrient content of the water, promoting a survival mechanism instigated by the bacteria.

The deterioration of water quality within potable water distribution systems is owing to a number of factors[30] including

- Bacterial colonisation of potable water pipe systems.
- Microbial regrowth.
- Poorly built and operated storage reservoirs.
- Taste and, in particular, odour problems owing to actinomyctes which produce an earthy smell, algae, and fungi.

Whilst potable water contains dissolved organic compounds, these can cause a number of problems including bad taste and odours, enhanced chlorine demand, trihalomethane formation, and bacterial colonisation of water distribution lines.[4,17,31-33]

7.4 RELEVANCE OF BIOFILMS IN POTABLE WATER

The depletion of dissolved oxygen, reduction of sulphates to hydrogen sulphide, bad taste, and colour may result from microbial activity within the sessile state. Problems of black water are sometimes caused by *Hyphomicrobium*, an example of which is the biofilm sloughing that occurred in Queensland, Australia.[34] The biofilm had accumulated on the pipe wall and contained high numbers of *Hyphomicrobium* which had formed hypha-like structures which are readily covered with a black precipitate of manganese oxide arising from low concentrations of reduced manganese in water. The cell wall of *Hyphomicrobium* acts like a catalyst in the oxidative precipitation process producing a black precipitate which would be transported in the water.

The occurrence of biofilms can cause many problems[22]

- Bacteria can be the starting point of a trophic food web leading to the proliferation of higher organisms.
- Specific bacterial species may generate turbidity, taste, and odours in the potable water.
- Production of red and black waters which result from the activity of both manganese and iron oxidising bacteria (*Hyphomicrobium*).
- High HPC interfere with detection of coliforms or sanitary indicators.
- Accumulation of attached biomass promotes corrosion, particularly under anaerobic conditions with the production of H_2S.
- Biofilms increase frictional resistance, thus decreasing the capacity of distribution systems to carry water.[35] This leads to a loss of pressure or reduced water flow. The increase in frictional resistance is known also to affect the microbial community in biofilms with an increase in filmanetous bacteria evident when this is high.[36]
- Continued failure of the distribution system to meet all established water quality criteria.

7.5 BIOFILM STRUCTURE IN POTABLE WATER

The structure of a biofilm is very difficult to study, particularly owing to the presence of large amounts of detritus, corrosion products, and inorganic matter, often scale (Figure 7.1), evident at the pipe surface in potable water. Evidence of these materials restricts the techniques which can be used to study the morphology of biofilms in potable water. Therefore, the majority of information which has become available in relation to biofilm structure in oligiotrophic environments, which are indicative of potable water environments, has been obtained from laboratory-based experiments.

Results on biofilm morphology suggests that, as far as potable water is concerned, disinfection processes cause effects on the biofilm structure. In general, it is found that in potable water that is chlorinated, biofilms have a physical appearance that is different from that of unchlorinated biofilms. In chlorinated water, compared to unchlorinated (control), bacterial cells tend to be clumped and the biofilm is found to be more patchy, with the cells approximately 50% smaller than the control cells.[37]

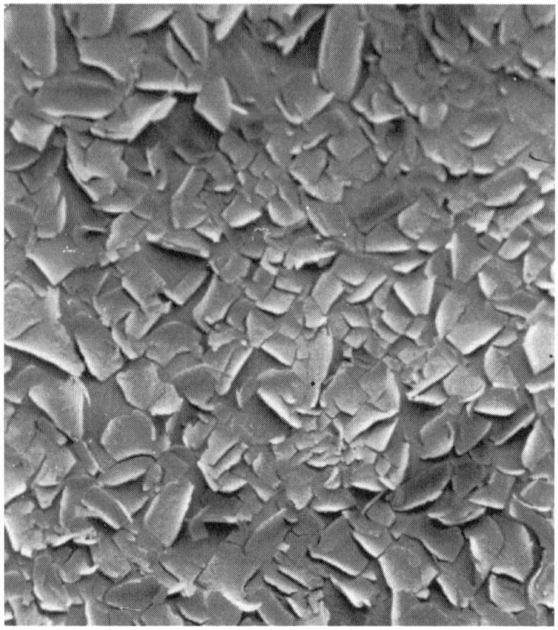

FIGURE 7.1 Evidence of scale in potable water pipes. Reprinted from *Water Research*, 32, Percival, S., Biofilms, mains water and stainless steel, 2187–2201, Copyright 1998, with permission from Elsevier Science.

There is also a shift from longer rod-shaped cells to more rounded organisms, together with a change in the dominant bacteria present in the biofilm.[37]

Biofilms were once considered a homogeneous growth of regular thickness.[26] However, direct examination of biofilms has demonstrated that bacteria grow in a number of microcolonies, the structure of which is dictated by physiological and environmental factors. For a critical review of the structure and function of biofilms, readers are advised to consult the review by Costerton et al.[38] This review draws together the historical and current literature to present biofilms as heterogeneous microniches present on a surface. As such, biofilms are now considered to consist of living, nonviable, or dead microorganisms together with extracellular polymeric substances, organic and inorganic matter in discrete microcolonies through which there are open water channels.[38-40] Biofilms in potable water systems are patchy (Figures 7.2 and 7.3) and heterogeneous with a diverse microbial flora (Figures 7.4).[41,42] A mature biofilm may have microcolony regions up to 200 μm thick. These regions will have differential concentration gradients which would have an effect on the physiology of the microcosm.[43,44] Owing to the thickness of microcolonies, the oxygen concentration would also vary depending on the depth into the microcolony. Hamilton[45] investigated the effect of oxygen on populations of sulphate-reducing bacteria whereas de Beer et al.[46] used microelectrodes to penetrate microcolonies. Anaerobic conditions within biofilms could lead to the growth of sulphate reducing bacteria whose activity may increase corrosion, particularly of the ductile iron once commonly used in water distribution systems.[26,47]

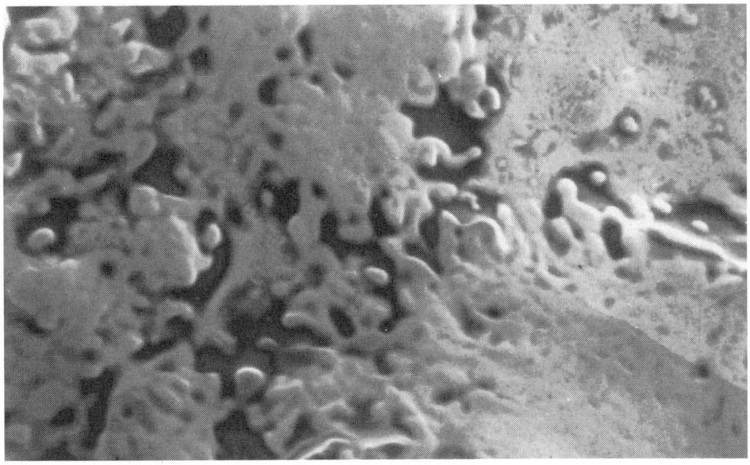

FIGURE 7.2 Evidence of a patchy biofilm formation in potable water. Reprinted from *Water Research,* 32, Percival, S., Biofilms, mains water and stainless steel, 2187–2201, Copyright 1998, with permission from Elsevier Science.

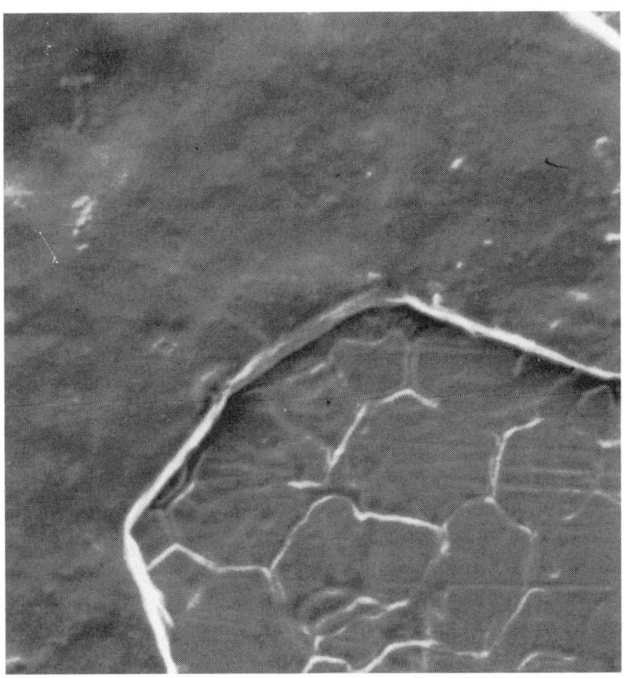

FIGURE 7.3 Evidence of a patchy biofilm formation in potable water. Reprinted from *Water Research,* 32, Percival, S., Biofilms, mains water and stainless steel, 2187–2201, Copyright 1998, with permission from Elsevier Science.

FIGURE 7.4 Evidence of a patchy heterogeneous biofilm. Reprinted from *Water Research,* 32, Percival, S., Biofilms, mains water and stainless steel, 2187–2201, Copyright 1998, with permission from Elsevier Science.

7.6 FACTORS WHICH AFFECT THE ACCUMULATION OF BIOFILMS IN POTABLE WATER

Within potable water, the main area of concern is that of proliferation of coliforms and heterotrophic bacteria in biofilms. Factors which are known to affect the growth of these include[37]

1. Utilisable carbon.
2. Temperature.
3. Ineffeciences in the removal/disinfection of organisms in treatment.
4. Disinfection dose/type.
5. The hydrodynamics of the distribution system.[48]

Whilst it is suggested that several parameters govern biofilm accumulation in water systems,[49] the most important variables to consider are particularly the water hydraulics, flux of cells, nutrients (AOC or BDOC), characteristics of the substratum, temperature, and disinfectants. The main parameters which are known to affect biofilm development within potable water will be discussed next.

7.6.1 Water Hydraulics

Water hydraulics have been instrumental in affecting microbial growth on pipe surfaces present in potable water. It is generally portrayed that high water velocities cause a greater flux of nutrients to a pipe surface, greater transport of disinfectants, and greater shearing of biofilms. Any changes in water velocity in potable water systems is owing mainly to seasonal fluctuations. Particular known causes are owing to consumer demand in the warm summers. Other factors known to affect water velocity include fire hydrants, pipe network design, breaks in pipes, or breaks in systems owing to pipe flushing.[50] Areas of pipe work which have stagnation zones are known to have adverse effects on biofilm in potable water. Particular effects include a loss in disinfectant residual and accumulation of sediment, debris, and, ultimately, increases in microbial growth.[51-53] The increase in microbial growth evident in these dead legs could then lead to high bacterial counts evident at the customer's taps.[54,55]

7.6.2 Nutrient Availability

The availability of nutrients is a necessity if bacteria are to grow in potable water. The general requirements for heterotrophic bacteria include carbon, nitrogen, and phosphorus in a ratio of about 100:10:1. This ratio is necessary for the maintenance of effective growth. Trace elements and cofactors, necessary chemicals for enzymes to function, are also required as well as these principal nutrients. The availability of these nutrients is one of the main factors known to affect the regrowth of bacteria within potable water which can be controlled during the water treatment processes.

Nutrients available to bacteria are generally in the form of AOC. This is often measured by the techniques initially utilised by van der Kooij, Visser, and Hijiner.[56] and have been used by a number of researchers in their studies[57] to demonstrate the influence of nutrients on growth. Production of water with a low AOC has been used within water distribution systems to reduce the total viable count and the presence of biofilms.[56] Other ways of measuring carbon in water include total organic carbon (TOC). This includes the total amount of soluble and insoluble organic carbon compounds, dissolved organic carbon (DOC), which includes the soluble fraction of TOC and AOC. This is the fraction of DOC that can be utilised for growth by microbes present in the water.

There is strong evidence available that suggests a correlation between AOC and regrowth of heterotrophic bacteria.[58] The potential threshold for assimilable organic carbon has been set at 10 mg of carbon per litre for heterotrophs[58] and 50 mg of carbon per litre for coliforms (LeChevallier, Schulz, and Lee, 1991).[59] It has also been found that AOC levels decrease with travel time through a potable water distribution system. With this in mind, it has been suggested that this decrease is owing to carbon utilisation by the sessile bacteria.[55] Despite evidence of a correlation between AOC levels and bacterial regrowth, particularly coliforms, other research has found no such evidence.[53]

Nitrogen, as well as carbon, is also considered a necessity for microbial life to develop amino acids and genetic material, and it is present in a number of different forms in water which can be utilised by bacteria. These include organic nitrogen, ammonia, nitrate, and nitrite. Despite the evidence of these in potable water, there is, at present, a lack of research and documented evidence on the effects of nitrogen levels on bacterial regrowth and biofilm formation. One piece of research by Donlan and Pipes[60] has indicated that no correlation exists between organic nitrogen, ammonia, nitrate, and nitrite, and bacterial growth in potable water. As with carbon, nitrogen is not presumed to be limiting in potable water, owing to the low concentrations which are necessary for bacterial growth. Phosphorus is also not presumed to be limiting in potable water. In the environment, phosphorus occurs as orthophosphate (PO_4^{3-}). Phosphate may be added to potable water to control corrosion. These have shown to have no real effect on the growth of bacteria.[61] However, some studies have found that low phosphorus concentrations may restrict microbial growth in potable water.[62]

7.6.3 Disinfection

Whilst chlorination is used to destroy potential pathogens and feacal coliforms in potable water, it is well known that increases of chlorine into potable water will not control biofilms or coliforms. Concentrations as high as 12 mg per litre have been shown to be inadequate to control biofilms in potable water.[63,64] It has also been found that the disinfection process can have profound effects on biofilm structure and morphology.[37] This suggests the development of a survival strategy by the biofilms caused by exposure to adverse conditions. The effects of disinfection on biofilm development will be discussed in Chapter 11.

7.6.4 Temperature

Temperature is well known to increase the microbial population within potable water systems with evidence of greater bacterial regrowth in potable water during summer months owing to higher water temperatures. Colbourne et al.[65] have demonstrated that coliform occurrence in potable water is associated with temperatures exceeding 20°C in comparison to a decrease when temperatures fell to less than 14°C. Donlan and Pipes[66] have also found a close relationship between water temperature and the development of biofilms. Increases in temperature above 15°C have led to general increases in the microbial content of potable water.[67-69] Fransolet, Villers, and Masschelein[68] have also found that water temperature not only has an effect on the growth rate of bacteria, but also the lag phase and cell yield. The length of lag time is important to the survival rate of bacteria in potable water. Temperature affects microbial growth rates, disinfection efficiency, dissipation of disinfectant residuals, corrosion rates, distribution system hydraulics, and water velocity through customer demand.[70] However, water companies have little control over changing water temperature. Therefore, any efforts should be focused on other parameters which temperature is known to affect. This can be reflected in the use of disinfectant residual adjustments.

7.6.5 Substratum Composition

Pipe material is well known to affect biofilm development in potable water. Different materials are used to transport potable water ranging from cast iron to unplasticised polyvinyl chlorides. A number of construction materials have been shown to stimulate bacterial growth. These include rubber, silicon, polyvinyl chloride (PVC), polyethylene, and bituminous coatings.[71-73] After several days of being immersed in potable water, it is possible to rank different materials used to transport potable water according to the number of attached cells: cast iron has more attached cells than cement lined cast iron which has more attached cells than stainless steel.[74] Plastic materials such as PVC and medium density polyethylene (MDPE) are in the process of replacing many of the older traditional materials such as cast iron for transporting potable water. However, like most materials which are exposed to potable water, their biofilm-forming potential has not been investigated to any large extent. Work completed recently has established that there are significant differences in terms of biofilm accumulation between cast iron, MDPE, and unplasticised PVC (uPVC) in drinking water.[75-77] Cast iron is noted in contributing to the deterioration of water quality and has been documented as being prone to microbial colonisation and, ultimately, microbial regrowth in potable water.[78-80]

It is possible to measure the biofilm formation potential (BFP) of materials which come into contact with potable water.[81] The test is based on measuring the biofilm density on materials which are incubated at 25°C in biostable drinking water as a function of time.[82] For glass, the BFP has been calculated at 10 pg ATP cm^{-2} compared to PVC (uPVC) of more than 100 pg ATP cm^{-2}.[82]

The relationship between the different surface characteristics and number of attached bacteria depends in particular on wettable area and the surface roughness of materials exposed to potable water.[7,83] Therefore, this suggests that any pipes used for the transport of potable water should be regarded as a potential health risk as the degree of surface roughness of piping systems ultimately will affect the adhesion of microorganisms.

Corrosion of water plumbing systems (particularly copper) of a few institutional buildings in certain soft water areas has been identified as being caused by the presence of biofilms.[57,84] This has led to the suggestion that alternative metals should be used where copper has shown to corrode prematurely. Corrosion provides a protective surface for microorganisms. Pits and nodules formed during the corrosion process are known to aid in the concentration of nutrients and to protect bacteria from water shear.[85] Coliforms have also been shown to be found in iron tubercles.[55] Although plastics would seem to be ideal materials of choice, some have been shown to impart nutrients, encouraging development of biofilms and pathogens, in particular, *Legionella pneumophila*.[75,86] One particular alternative to plastics, iron, and steel pipes for transporting potable water that is now being extensively researched is stainless steel.[6,7] Overall, the use of materials that will reduce biofilm development, owing to their own characteristics, will obviously be advantageous in potable water and thus prevent/reduce the public health potential which can exist in biofilms.

7.7 COLIFORM REGROWTH IN POTABLE WATER

Coliforms are used as indicator organisms in potable water to confirm the presence of faecal contamination. If they are present in potable water, they are generally evident owing to a number of reasons but, in particular, disinfection failures. In the U.S., it has been documented that of 164 water utilities surveyed, a high number of these experience recurring coliform episodes.[87] These have been owing mainly to disinfection barrier breakthroughs.

Despite evidence of coliform regrowth in potable water, there is still little evidence that conclusively supports coliform proliferation on pipe walls. This has been owing to a number of reasons relating predominantly to the difficulty of sampling potable water pipe systems and the heterogeneity of biofilms known to be evident in potable water. Contrary to this, a large number of researchers have found evidence of coliform regrowth and consider this a public health concern.

7.8 PROBLEMS OF STUDYING POTABLE WATER BIOFILMS IN EXPERIMENTAL MODELS

The study of biofilms within potable water distribution systems and mimicking the process in laboratory-based systems are difficult. This is because these systems are chaotic and fluctuate continually, owing to the large number of variable changes known to be present at any moment in time. These range from sudden increases in temperature which, in turn, lead to increases in turbidity owing to increasing nutrient levels. Even more difficult to model are the hydraulic regimes and constantly fluctuating pressure and flow conditions in water distribution systems together with changes and/or hot–low spots of disinfectants.

As Camper[88] portrayed, there are two basic types of pilot systems developed to study biofilms. These are once through systems[7] and recirculating systems.[7] Possibly, the most extensively researched study of biofilms in potable water have taken place using laboratory-based simulations and large scale pilot rig systems.[14,18,89,90] Possibly, the largest was undertaken at Thames Water, U.K.[91]

7.9 CONCLUSION

Potable water biofilms are not well characterised. It should be presumed that all surfaces within potable water systems have a biofilm presence. By reviewing the literature, it is possible to draw a number of conclusions with respect to biofilm formation in potable water. It is evident that organic carbon is by far the most important nutrient source able to support the growth of sessile communities which can range from opportunistic and pathogenic bacteria to normal noninfectious heterotrophic plate count bacteria. This has been shown to be the case in laboratory-based studies using pilot and full scale rig systems. It is also now acknowledged that temperature, linked closely to nutrient and, therefore, turbidity levels, has important effects on biofilm formation in potable water. However, with this in mind, very little work has been done in this area. It is, nevertheless, feasible to suggest that increasing temperature leads to increasing biofilm growth up to an optimum and has pronounced effects on disinfection efficiency.

7.10 REFERENCES

1. Committee on Water Supply, 1930, Bacterial aftergrowths in distribution systems, *Am. J. Public Health,* 20, 485.
2. Baylis, J. R., 1930, Bacterial aftergrowths in water distribution systems, *Water Works Sewerage,* October, 335.
3. Adams, G. O. and Kingsbury, F. H., 1937, Experiences with chlorinating new water mains, *J. New Engl. Water Works Assoc.,* 60.
4. Allen, M. J., Taylor, R. H. and Geldreich, E. E., 1980, The occurrence of microorganisms in water main encrustations, *J. Am. Water Works Assoc.,* 72, 614.
5. Ridgway, H. F. and Olson, B. H., 1981, Scanning electron microscope evidence for bacterial colonization of a drinking-water distribution system, *Appl. Environ. Microbiol.,* 41, 274.
6. Percival, S. L., Knapp, J. S., Wales, D. S., and Edyvean, R., 1998, Biofilm development on stainless steel in mains water, *Water Res.,* 32, 243.
7. Percival, S. L., Beech, I. B., Knapp, J. S., Edyvean, R., and Wales, D. S., 1997, Biofilm development on stainless steels in a potable water system, *J. Inst. Water Environ. Manage.,* 11, 289.
8. van der Kooij, D., 1977, The occurrence of *Pseudomonas* spp. in surface water and tap water as determined on citrate media, *Antonie van Leeuwenhoek,* 43, 187.
9. Burman, N. P., 1973, The occurrence and significance of actinomycetes in water supply, in *Actinomycetaeles: Characteristics and Practical Importance,* Sykes, G. and Skinner, F. A., Eds., Academic Press, London, New York, 219.
10. Olson, B. H., 1982, Assessment and implications of bacterial regrowth in water distribution systems, Project Report EPA R-805680010, Order No. PB 82-249368, National Technical Information Service, Spingfield, VA.
11. Lee. S. H., O'Connor, J. T., and Banerji, S. K., 1980, Biologically mediated corrosion and its effects on water quality in distribution systems, *J. Am. Water Works Assoc.,* 72, 636.
12. Wagner, D., Fischer, W., Walker, J. T., and Keevil, C. W., 1994, Rapid detection of biofilm on corroded copper pipes, *Biofouling,* 8, 47.
13. Block, J. C., 1992, Biofilms in potable water distribution systems, in *Biofilms: Science and Technology,* Melo, L. F., Bott, T. R., Fletcher, M., and Capdeville, B., Eds., May 18–29, Alvor, Portugal, Kluwer Academic Publishers, London, 469.
14. Mathieu, L., Paquin, J. L., Block, J. C., Randon, G., Maillard, G., and Reasoner, D. J., 1992, Parameters governing bacterial regrowth in distribution systems, *J. Fr. Hydrol.,* 15, 207.
15. Block, J. C., Mathieu, L., Servais, P., Fontvielle, D., and Werner, P., 1992, Indigenous bacterial inocula for measuring the biodegradable dissolved organic carbon (BDOC) in waters, *Water Res.,* 26, 481.
16. LeChevallier, M. W., Becker, W. C., Schorr, P., and Lee, R. G., 1992, Evaluating the performance of biologically active rapid filters, *J. Am. Water Works Assoc.*, 84, 136.
17. Servais, P., Billen, G., Ventresque, C., and Bablon, G. P., 1991, Microbial activity in GAC filters at the Choisy–le Roi treatment plant, *J. Am. Water Works Assoc.*, 83, 62.
18. Block, J. C., Haudidier, K., Paquin, J. L., Miazga, J., and Levi, Y., 1993, Biofilm accumulation in drinking water distribution systems, *Biofouling,* 6, 333.

19. Jess, J., Hight, N. J. Eptoin, H. A. S., Sigee, D. C., O'Neil, J. G., Meier, H., Handley, P. S., 1997, Comparison of the API 20NE biochemical identification systems and partial 16s rRNA gene sequencing for the characterisation of culturable heterotrophs from a water distribution system biofilm, in *Biofilms: Community Interactions and Control*, Wimpenny, J., Handley, P., Gilbert, P., Lappin-Scott, H., and Jones, M., Eds., Bioline, Cardiff, 149.
20. O'Conner, J. T. and Banerji, S. K., 1984, Biologically mediated corrosion and water quality deterioration in distribution systems, EPA Report 600/2/84/056, Cincinnati, OH, 442.
21. Olson, B. H. and Nagy, L. A., 1984, Microbiology of potable water, *Adv. Appl. Microbiol.*, 30, 73.
22. Fransolet, G., Villers, G., Goyens, A., and Masschelein, W. J., 1986, Metabolic identification of germs identified from ozonized water mixed with underground water, *Ozone Sci. Eng.*, 8, 95.
23. Maki, J. S., La Croix, S. J., Hopkins, B. S., and Staley, J .T., 1986, Recovery and diversity of heterotrophic bacteria from chlorinated potable waters, *Appl. Environ. Microbiol.*, 51, 1047.
24. Scarpino, P. V., Kellner, G. R., and Cook, H. C., 1987, The bacterial quality of bottled water, *J. Environ. Sci. Health,* A22, 357.
25. McFeters, G. A., Camper, A. K., LeChevallier, M. W., Broadway, S., and Davies, D. G., 1987, Bacteria attached to granular activated carbon in potable water, EPA Environmental Research Brief, June 1–5, 1987, Cincinatti, OH.
26. Hamilton, W. A., 1985, Sulphate-reducing bacteria and anaerobic corrosion, *Ann. Rev. Microbiol.*, 39, 195.
27. LeChevallier, M. W., Babcock, T. M., and Lee, R. G., 1990, Disinfecting biofilms in a distribution system, *J. Am. Water Works Assoc.* 82, 87.
28. Paquin, J. L, Block, J. C., Haudidier, K., Hartemann, P., Colin, F., Miazga, J., and Levi, Y., 1992, Effect of chlorine on the bacterial colonisation of a model distribution system, *Rev. Sci. DE L'Eau,* 5, 399.
29. LeChevallier M. W, Cawthon, C. D., and Lee, R. G., 1988, Inactivation of biofilm bacteria, *Appl. Environ. Microbiol.,* 54, 2492.
30. Sobsey, M. D. and Olson, B., 1983, Microbial agents of waterborne disease, in *Assessment of Microbiology and Turbidity Standards for Drinking Water*, Berger, P. S. and Argaman, Y., Eds., EPA-570-9-83-001, Office of Drinking Water, Washington, D.C.
31. Bourbigot, M. M., Dodin, A., and Lheritier, R., 1984, Bacteria in distribution systems, *Water Res.*, 18, 585.
32. Hoehn, R. C., Barnes, D. B., Thompson, B. C., Randall, C. W., Grizzard, T. J., and Shaffer, P. T. B., 1980, Algae as sources of trihalomethane precursors, *J. Am. Water Works Assoc.*, 72, 344.
33. Rook, J. J., 1974, Formation of haloforms during chlorination of natural waters, *Water Treat. Exam.*, 23, 234.
34. Hamilton, W. A. and Maxwell, S., 1986, Biological and corrosion activities of sulphate-reducing bacteria within natural biofilms, in *Biologically Induced Corrosion,* Dexter, S. C., Ed., National Association of Corrosion Engineers, Houston, TX, 131.
35. Bryers, J. D. and Characklis, W. G., 1981, Early fouling biofilm formation in a tubular flow system: overall kinetics, *Water Res.,* 15, 483.
36. Trulear, M. G. and Characklis, W. G., 1982, Dynamics of biofilm processes, *J. Water Pollut. Control Fed.,* 54, 1288.

37. Camper, A., Burr, M., Butterfield, P., and Abernathy, C., 1999, Development and structure of drinking water biofilms and techniques for their study, *J. Appl. Microbiol. Symp. Suppl.,* 85, 1S.
38. Costerton, J. W., Lewandowski, Z., DeBeer, D., Caldwell, D., Korber, D., and James, G., 1994, Biofilms, the customised microniche, *J. Bacteriol.,* 176, 2137.
39. Keevil, C. W., and Walker, J. T., 1992, Normarski D.I.C. microscopy and image analysis of biofilms, *Binary,* 4, 93.
40. Sutherland, I., 1997, Microbial biofilms; exopolysaccharides – superglue or velcro?, in *Biofilms: Community Interactions and Control,* Wimpenny, J., Handley, P., Gilbert, P., Lappin-Scott, H. M., and Jones, M., Eds., Bioline, Cardiff, 39.
41. Lawrence, J. R., Korber, D. R., Hoyle, B. D., Costerton, J. W., and Caldwell, D. E., 1991, Optical sectioning of microbial biofilms, *J. Bactiol.,* 173, 6558.
42. Keevil, C. W., Rogers, J., and Walker, J. T., 1995, Potable water biofilms, *Microbiol. Eur.,* 3, 10.
43. Rodrguez, G. G., Phipps, D., Ishiguro, K., and Ridgway, H. F., 1992, Use of a fluorescent redox probe for direct visualisation of actively respiring bacteria, *Appl. Environ. Microbiol.,* 58, 1801.
44. Xu, K. D., Stewart, P. S., Xia Huang, C. T., and McFeters, G. A., 1998, Spatial physiological heterogeneity in *Pseudomonas aeruginosa* biofilm is determined by oxygen availability, *Appl. Environ. Microbiol.,* 64, 4035.
45. Hamilton, W. A., 1983, Sulphate-reducing bacteria and the offshore oil industry, *Trends Biotechnol.,* 1, 36.
46. de Beer, D., Stoodley, P., Roe, F., and Lewandowski, Z., 1994, Effects of biofilm structures on oxygen distribution and mass transfer, *Biotechnol. Bioeng.,* 43, 1131.
47. Santegoeds, C. M., Ferdelman, T. G., Muyzer, G., and de Beer, D., 1998, Structural and functional dynamics of sulfate-reducing populations in bacterial biofilms, *Appl. Environ. Microbiol.,* 64, 3731.
48. Stoodley, P., Dodds, I., Boyle, J. D., and Lappin-Scott, H. M., 1999, Influence of hydrodynamics and nutrients on biofilm structure, *J. Microbiol. Symp. Suppl.,* 85, 19S.
49. Bryers, J. D., 1987, Biologically active surfaces: processes governing the formation and persistence of biofilms, *Biotechnol. Bioeng.,* 24, 2451.
50. Geldreich, E. E., 1988, Coliform noncompliance nightmares in water supply distribution systems, in *Water Quality: A Realistic Perspective*, Univ. Michigan, College of Engineering; Michigan Section, American Water Works Association; Michigan Water Pollution Control Association; Michigan Dept. of Public Health, Lansing, MI, chap. 3.
51. Smith, D. B., Hess, A. F., and Opheim, D., 1989, Control of distribution system coliform regrowth, *Proc. AWWA Water Quality Technol. Conf.,* Philadelphia, PA, American Water Works Association, Denver, CO.
52. Hanson, H. F., Mueller, L. M., Hasted, S. S., and Goff, D. R., 1987, Deterioration of water quality in distribution systems, American Water Works Association, Denver, CO.
53. Opheim, D. and Smith, D., 1990, The relationship of *Enterobacter cloacae*, AOC, and ADOC in raw and treated water to confirm episodes, in *Proc. AWWA Water Quality Technol. Conf., San Diego, CA,* American Water Works Association Research Foundation, Denver, CO, 1237.
54. Brazos, B. J., O'Connor, J. T., and Abcouwer, S., 1985, Kinetics of chlorine depletion and microbial growth in household plumbing systems, *Proc. AWWA Water Quality Technol. Conf.,* Houston, TX.

55. LeChevallier, M. W., Babcock, T. M., and Lee, R. G., 1987, Examination and characterization of distribution system biofilms, *Appl. Environ. Microbiol.,* 53, 2714.
56. van der Kooij, D., Visser, A., and Hijiner, W. A. M., 1982, Determining the concentration of easily assimilable organic carbon in drinking water, *J. Am. Water Works Assoc.,* 74, 540.
57. Keevil, C. W., Walker, J. T., McEvoy, J., and Colbourne, J. S., 1989, Detection of biofilms associated with pitting corrosion of copper pipework in Scottish hospitals, in *Biocorrosion,* Gaylarde, L. C. and Morton, L. H. E., Eds., Biodeterioration Society, Kew, England, 99.
58. van der Kooij, D., 1992, Assimilable organic carbon as an indicator of bacterial regrowth, *J. Am. Water Works Assoc.,* 84, 57.
59. LeChevallier, M. W., Schulz, W. H., and Lee, R. G., 1991, *Giardia* and *Cryptosporidium spp.* in filtered water supplies, *Appl. Environ. Microbiol.,* 57, 2617.
60. Donlan, R. M. and Pipes, W. O., 1988, Selected drinking water characteristics and attached microbial population density, *J. Am. Water Works Assoc.*, 80, 70.
61. Rosenzweig, W. D., 1988, Influence of phosphate corrosion control compounds on bacterial growth, EPA Project Summary CR-811613-01-0, Environmental Protection Agency, Cincinnati, OH.
62. Herson, D. S., Marshall, D. R., and Victoreen, H. T., 1984, Bacterial persistence in the distribution system, in *Proc. AWWA Water Quality Technol. Conf., Denver, CO,* American Water Works Association Research Foundation, Denver, CO, 309.
63. Earnhardt, K. B., 1980, Chlorine resistant coliforms — the Muncie, Indiana experience, in *Proc. AWWA Water Quality Technol. Conf., Miami Beach, FL,* American Water Works Association Research Foundation, Denver, CO, 371.
64. Lowther, E. D. and Moser, R. H., 1984, Detecting and eliminating coliform regrowth, in *Proc. AWWA Water Quality Technol. Conf., Denver, CO,* American Water Works Association Research Foundation, Denver, CO, 323.
65. Colbourne, J. S., Dennis, P. J., Keevil, W., and Mackerness, C., 1991, The operational impact of growth of coliforms in London's distribution system, in *Proc. AWWA Water Quality Technol. Conf., Orlando, FL,* American Water Works Association Research Foundation, Denver, CO, 799.
66. Donlan, R. M. and Pipes, W. O., 1986, Pipewall biofilm in potable water mains, 14th Annual AWWA Water Quality Technology Conference, Portland, OR.
67. Rizet, M., Fiessinger, F., and Houel, N., 1982, Bacterial regrowth in a distribution system and its relationship with the quality of the feed water: case studies, *Proc. AWWA Annual Conf., Miami Beach, FL,* May 16–20.
68. Fransolet, G., Villers, G., and Masschelein, W. J., 1985, Influence of temperature on bacterial development in waters, *J. Int. Ozone Assoc.*, 7, 205.
69. LeChevallier, M. W., 1990, Coliform regrowth in drinking water: a review, *J. Am Water Works Assoc.,* 82, 74.
70. **Anon.,** 1992, Control of biofilm growth in drinking water distribution systems, Seminar Publication, Environmental Protection Agency, Office of Research and Development, Washington, D.C.
71. Schoenen, D. and Scholer, H. F., 1985, *Drinking Water Materials: Field Observations and Methods of Investigation,* John Wiley & Sons, New York.
72. Frensch, K., Hahn, J. U., Levsen, K., Nieben, J., Scholer, H. F., and Schoenen, D., 1987, Solvents from the coating of a storage tank as a reason of colony increase in drinking water, *Vom Wasser,* 68, 101.
73. Schoenen, D. and Wehse, A., 1988, Microbial colonization of water by the materials of pipes and hoses: changes in colony counts, *Zentralbl. Bakt. Hyg. B*, 186, 108.

74. Block, J. C., Paquin, J. L., Colin, F., Miazga, J., and Levi, Y., 1991, Accumlation of bacteria on different types of material: a study comparing distribution, Seminar on Adhesion of Microorganisms to Different Solid Surfaces, Massey, IN.
75. Rogers, J., Dowsett, A. B., Dennis, P. J., Lee, J. V., and Keevil, C. W., 1994, Influence of plumbing materials on biofilm formation and growth of *Legionella pneumophila* in potable water systems, *Appl. Environ. Microbiol.,* 60, 1179.
76. Holden, B., Greetham, M., Croll, B. T., and Scutt, J., 1995, The effect of changing inter process and final disinfection reagents on corrosion and biofilm growth in distribution pipes, *Water Sci. Technol.,* 32, 213.
77. Kerr, C. J., Osborn, K. S., Robson, G. D., and Handley, P. S., 1999, The relationship between pipe material and biofilm formation in a laboratory model system, *J. Appl. Microbiol. Symp. Suppl.,* 85, 29S.
78. Camper, A. K., 1996, Factors limiting microbial growth in the distribution System: Piolet and laboratory studies, American Water Works Association Research Foundation, Denver, CO.
79. Camper, A. K., Jones, W. L., and Hayes, J. T., 1996, Effect of growth conditions and substratum composition on the persistence of coliforms in mixed population biofilms, *Appl. Environ. Microbiol.,* 62, 4014.
80. LeChevallier, M. W., Welch, N. J., and Smith, D. B., 1996, Factors limiting microbial growth in the distribution system: full scale experiments, American Water Works Association Research Foundation, Denver, CO.
81. van der Kooij, D. and Veenendaal, H. R., 1994, Assessment of the biofilm formation potential of synthetic materials in contact with drinking water during distribution, in *Proc. AWWA Water Quality Technol. Conf., Miami, FL, 1993,* American Water Works Association, Denver, CO, 1395.
82. van der Kooij, D., 1999, Potential for biofilm development in drinking water distribution systems, *J. Appl. Microbiol. Symp. Suppl.,* 85, 39S.
83. Korber, D. R., Choi, A., Wolfaardt, G. M., Ingham, S. C., and Caldwell, D. E., 1997, Substratum topography influences susceptibility of *Salmonella enteritidis* biofilms to trisodium phosphate, *Appl. Environ. Microbiol.,* 63, 3352.
84. Geesey, G. G., Bremer, P. J., Fischer, W. R., Wagner, D., Keevil, C. W., Walker, J. T., Chamberlain, A. H. L., and Angell, P., 1994, Unusual types of pitting corrosion of copper tubes in potable water systems, in *Microbial Biofilms,* Geesey, G. G., Ed., Lewis Publishing, New York, 243.
85. Allen, M. J., Taylor, R. H., and Geldreich, E. E., 1980, The occurrence of microorganisms in water main encrustations, *J. Am. Water Works Assoc.,* 72, 614.
86. Bezanson, G., Burbridge, S., Haldance, D., and Marrie, T., 1992, *In situ* colonisation of polvinyl chloride, brass and copper by *Legionella pneumophila, Can. J. Microbiol.,* 38, 3322.
87. Smith, D. B., Hess, A. F., and Hubbs, S. A., 1990, Survey of distribution system coliform occurrence in the United States, in *Proc. AWWA Water Quality Technol. Conf., San Diego, CA,* American Water Works Association Research Association, Denver, CO, 1103.
88. Camper, A. K., 1994, Coliform regrowth and biofilm accumulation in drinking water systems: a review, in *Biofouling and Biocorrosion in Industrial Water Systems,* Geesey, G. G., Lewandowski, Z., and Flemming, H. C., Eds., CRC Press, Boca Raton, FL, 91.
89. LeChevallier, M. W., Schulz, W. H., and Lee, R. G., 1990, Bacterial nutrients in drinking water, in *Assessing and Controlling Bacterial Regrowth in Distribution Systems,* American Water Works Association and American Water Works Association Research Foundation, Denver, CO.

90. Mackerness, C. W., Colbourne, J. C., Dennis, P. J., Rachwal, A. J., and Keevil, C. W., 1993, Formation and control of coliform biofilms in drinking water distribution systems, Soc. Applied Bacteriology Technical Series, Microbial Biofilms, 217.
91. Holt, D., 1995, Challenge of controlling biofilms in water distribution systems, in *The Life and Death of Biofilms,* Wimpenny, J., Handley, P., Gilbert, P., and Lappin-Scott, H., Eds., Bioline, Cardiff, 161.

8 Microbes and Public Health Significance

CONTENTS

8.1 MICROBIOLOGY OF POTABLE WATER SYSTEMS

Potable water of good quality is a necessity in reducing incidences of waterborne disease outbreaks, and preventing the deterioration of public health. Developments in water treatment practices have been paramount in reducing these incidences, particularly in developing countries. However, treatment methods used to remove pathogenic microbes have not been completely successful at reducing all incidences of waterborne disease outbreaks within potable water supplies. The global effect of waterborne diseases is colossal, with more than 250 million new cases being reported each year, resulting in over 10 million deaths.[1] On a worldwide scale, the most commonly encountered waterborne diseases are caused by *Salmonella, Shigella, Escherichia coli, Campylobacter* spp., and *Vibrio cholerae*.

8.2 UNTREATED WATER

Untreated water contains many bacteria, protozoa, fungi, and viruses. Constituting parts of these groups are primary pathogens and opportunistic pathogens such as *Acinetobacter* spp., *Aeromonas* spp., *Flavobacterium* spp., *Moraxella* sp., *Corynebacterium* spp., *Arthrobacter* spp., and *Pseudomonas* spp.[2] Whilst water treatment methods are intended to remove or inactivate pathogenic microorganisms from untreated water, these opportunistic pathogens, in particular, *Aeromonas hydrophila* and *Pseudomonas aeruginosa*, can be detected within filtration plants at water treatment facilities which may pass into the normal potable water supply system.[2]

Owing to the potential risk of waterborne diseases, microbial contamination is considered to be the most critical risk factor in maintaining good potable water quality.[3] Many diseases which are attributable to microbes associated with potable water include gastroenteritis, giardiasis, amoebiasis, cryptosporidiosis, shigellosis, and hepatitis, salmonellosis, and, of growing concern in developing countries, cholera. The World Health Organisation (WHO) calculated that in 1994 there were up to 5.5 million cases of cholera annually, with up to 120,000 deaths[4] being reported in Asia and Africa.

8.3 TREATED WATER SYSTEMS

Ultimately, control of water systems with pathogens and potentially pathogenic bacteria involves treatment with the use of disinfectants. The efficiency of particular biocides on pathogenic bacteria is generally obtained by its effective biocidal activity on indicator organisms, in particular, the coliform group. This is a general approach adopted by the WHO,[5] U.S. Environmental Protection Agency (EPA),[6] and European Union (EU).[7] However, waterborne diseases are no longer specific to just enteric bacteria but are caused by a broader spectrum of microbes not necessarily monitored for within a potable water system.[8] This has been owing to the fact that reports of waterborne outbreaks have largely highlighted the effects of protozoan and viral agents in water considered safe to drink under current legislation.[8] The major infectious agents which are found in contaminated drinking water worldwide are shown in Table 8.1.

Associated with higher consumer expectations for quality water supplies are the effects of chlorine and its associated trihalomethanes (THM) on human health because they are well known to be both toxic and carcinogenic in high levels. Whilst pathogenic microbes are generally absent from potable water supplies owing to the

TABLE 8.1
Major Infectious Agents Found in Contaminated Drinking Water Worldwide

Bacteria	Virus	Protozoa	Helminths
Campylobacter jejuni	Adenovirus (31 types)	*Balantidium coli*	*Ancylostoma duodenale*
Enteropathogenic *E. coli*	Enterovirus (71 types)	*Entamoeba histolytica*	*Ascaris lumbricoides*
Salmonella (170 spp.)	Hepatitis A	*Giardia lamblia*	*Echinococcus granulosis*
Shigella (4 spp.)	Norwalk virus		*Necator americanus*
Cryptosporidium	Reovirus		*Strongyloides stercoralis*
Vibrio cholera	Rotavirus		*Taenia* sp.
Yersinia enterocolitica	Coxsackie virus		*Trichuris trichiurs*

Source: Adapted from Geldreich.[116]

usage of chlorine, contamination by pathogens can occur, particularly owing to unsatisfactory repairs or leakage in plumbing or distribution systems. This would ultimately generate public health worries with respect to enteric microbial presence and also to an increased dosage of chlorine for their removal.

As numbers of microorganisms of potentially public health significance are present in potable water, they have the possibility of adhering to the pipes present in water mains and developing into a biofilm. [9-11]

8.4 CONTAMINATION SOURCES OF POTABLE WATER

Suppliers of potable water aim to provide water "free of pathogens" with no intention of providing a sterile product. Therefore, there are a number of possible sources of contamination in a potable water system including:

- At source, particularly with respect to reservoirs: although contamination could occur within the potable water system. It is generally found that the majority of diseases detected in potable water are owing to untreated or inadequately treated groundwater.[12]
- Water treatment plants: these will never eliminate all microorganisms. Whilst most bacteria will be killed after disinfection, there will still be a small population remaining either intact or uninjured.
- Within the distribution system: water may undergo deterioration before or when it reaches the consumer's tap either by the disinfectant residual, hydraulic retention time, flow rates, pipe material, temperature, source of water and corrosion leading to a supply of nutrients, or the development of a biofilm.[13] The introduction of pathogens into the distribution systems could be owing to fractures in the water mains leading to contamination of the potable water.[14]
- Biofilm development: as water passes through a potable water system, the microbiological properties of the water may be altered by the presence of a biofilm.

Outbreaks of waterborne illness caused by cross-connections, back-siphonage, or contaminated potable water is shown in Table 8.2.

8.5 THE PUBLIC HEALTH SIGNIFICANCE OF BIOFILMS IN POTABLE WATER

Over the last decade there have been a number of incidents of waterborne outbreaks in waters that have met the required legal standards for faecal indicator organisms. This could be possibly owing to improved methods of reporting such outbreaks, improved detection methods, improvements in surveillance systems, the protective nature of biofilms, or poor biocidal activity in problem areas.

TABLE 8.2
Outbreaks of Waterborne Illness Caused by Cross-Connections, Back-Siphonage, or Contaminated Water Mains, 1920–1984

Diseases	Number of Outbreaks Associated with	
	Cross-Connections Back Siphonage	Contamination of Water Mains
Gastroenteritis, etiology unknown	74	19
Typhoid fever	37	6
Chemical poisoning	14	0
Hepatitis A	8	4
Shigellosis	6	1
Giardiasis	4	2
Salmonellosis	3	1
Viral gastroenteritis	2	1
Amoebiasis	0	1
Other	2	0
Total	150	35

Source: Adapted from Geldreich.[116]

Biofilms are known to harbour large numbers of microorganisms that could remain undetected until they are sloughed off by possible water shear. This suggests biofilm growth and that its microbial ecology should be monitored, together with the microbiology of potable water, to a far greater degree.

Many waterborne outbreaks which have been shown to be on the increase in potable water have been owing to mainly protozoan parasites (*Giardia* and *Cryptosporidium*) and enteric viruses. Growing concern is also developing with respect to microbes like *Legionella* spp., *Helicobacter pylori*, *Mycobacterium* spp., and *Aeromonas* spp. together with other pathogenic and opportunistic pathogens. Despite evidence of the existence of biofilms in potable water systems, very little information is available on the occurrence of coliforms and pathogenic bacteria in this area. This suggests that more research is necessary to obtain data of waterborne incidences associated with biofilms.

Potable water quality is paramount in the spread of disease. Studies on water quality have highlighted a large species diversity of bacteria. Bacteria associated with potable water which have been identified as problematic but generally not classed as harmful include *Pseudomonas* spp., *Flavobacterium* sp., *Spirillium* sp., *Clostridium* sp., *Arthrobacter* sp., *Gallionella*, and *Lepthothrix* spp. However, *Flavobacterium* and *Pseudomonas* spp. are well known opportunistic pathogens with *Pseudomonas aeruginosa* being a major cause of hospital infections.

Other bacteria of major concern include *Campylobacter* sp. and, in particular, viral gastroenteritis which has been found to be associated with 66% of waterborne disease outbreaks in the U.K. between 1977 to 1986,[15] and *Cryptosporidium* which has been shown to constitute 64% of 11 waterborne outbreaks detected in the U.K.

between 1987 to 1990.[16] Together with the previously named microbes, the most commonly identified bacteria in most waterborne outbreaks in the U.S. have been *Salmonella* and *Shigella*. Despite high incidences of outbreaks associated with microorganisms, it has been estimated that only 50% of waterborne diseases in municipal water systems have been reported.

As previously outlined, opportunistic pathogens associated with potable water systems and biofilms are slowly emerging as a serious public health threat. They form part of the normal heterotrophic flora detected in potable water systems.[17] Whilst opportunistic pathogens are not generally classed as a potential health threat in certain situations, they have been shown to lead to infections in the elderly, immunocomprised patients, and babies.[18]

There is an underlying problem in assessing the health implications of organisms which reside both in potable water and within biofilms[8] because there is a lack of information on the

- Densities and occurrence of these organisms.
- Infection dose data required to establish an infection.
- Incidence of diseases caused by exposure to such organisms.
- Effectiveness of treatment procedures and post-disinfection for control of these agents particularly in biofilms.
- Availability of good detection methods and surveillance systems to monitor these organisms both in the planktonic and sessile states.

There has been a recent increase in research in the regrowth of biofilm and nutrients in potable water. Some studies have involved laboratory simulation models,[19] closed loop systems,[20] small continuous flow through pipe-based systems,[13,21] or large pilot scale network systems to simulate distribution systems.[22-24]

8.6 MICROORGANISMS OF CONCERN WHICH ARE ASSOCIATED WITH POTABLE WATER AND BIOFILMS

Potable water biofilms are dominated by Gram-negative bacteria[25-27] particularly opportunistic pathogens such as *Pseudomonas* and *Flavobacterium*-like species.[28-33] As well as regular harmless flora, pathogenic or potentially pathogenic organisms may be present (e.g., *Legionella*), which could have serious consequences when detected as part of a biofilm and dispersed as an aerosol.[34] The heterotrophic bacteria that make up the basic water-distribution biofilm have been poorly characterised owing mainly to the inaccuracy of standard available identification kits. However, with the development of molecular techniques, identification is proving to be more accurate.[35,36]

The presence of *Cryptosporidium*, *Campylobacter jejuni*, *Legionella*, *Giardia*, *Aeromonas* sp., *Mycobacteria* sp., *Salmonella* sp., *Helicobacter pylori*, *Shigella* spp., and viruses are causing concern in both potable water and biofilms. Each of these will be considered in turn in this chapter. It may be slightly misleading to include *Cryptosporidium*, *Giardia*, and viruses in the list of biofilm-associated organisms.

Whilst it is true that if laboratory model systems are seeded with very high numbers of these organisms, they can be detected subsequently for weeks in the biofilm, but unlike bacteria they do not have the capability to proliferate in such environments.[25] In water distribution systems, contamination with the above microorganisms may pose a significant health risk.[19,23]

In a number of epidemiological studies it has been observed that by installing reverse osmosis filters in households, the levels of gastrointestinal illness can be reduced by some 35%.[37] Other work carried out by Payment[38] has found that despite potable water meeting the required legal standards, it still remains a source of a significant level of potable water attributable illnesses.

The role of biofilms in protecting both nonpathogenic and pathogenic organisms is not well characterised, particularly from disinfectant residuals presently being used, and in some cases enhancing their ability to amplify within the distribution system. Therefore, it is very possible that biofilms protect, enhance, and/or promote the growth of indigenous pathogens in general. Also, other questions that remain unanswered include: what is the relative role of the ecology of the distribution systems and does this ecology promote the survival and growth of various pathogens evident within a biofilm? This, therefore, suggests that there may be a public health significance owing to the development of biofilms within potable water, particularly in light of this emergence of existing, new, and uncharacterised pathogens.

Despite evidence of the existence of biofilms in potable water systems, very little information is available on the occurrence of coliforms and pathogenic bacteria in this area. This chapter looks at potential microbes that do and could become a significant public health threat if accumulated in biofilms and sloughing takes place both randomly and routinely.

8.7 BACTERIA

8.7.1 General Characteristics and Association with Potable Water

Bacteria constitute the largest proportion of the sessile and planktonic populations in potable water. These can be measured by the heterotrophic plate count (HPC). The predominant HPC bacteria found in potable water and biofilms are shown in Tables 8.3. As these represent the largest proportion of bacteria identified within biofilms, they must be able to survive water treatment processes in order for them to colonise pipe surfaces. They could also be introduced into potable water by a number of other means. These include cross-connections, backflow events, pipe leakage, or even repair operations.

The HPC is intended to detect gram-negative and gram-positive organisms, namely, *Pseudomonas*, *Aeromonas*, *Klebsiella*, *Flavobacterium*, *Enterobacter*, *Citrobacter*, *Alcaligenes*, and *Moraxella* with gram-positive bacteria such as *Bacillus* and *Micrococcus* also being found.[39] A number of these HPC bacteria species are known to be antagonistic to the coliform growth and often lead to pathogen protection.[37,40] Known bacteria which show evidence of this include *Pseudomonas* spp., *Sarcina* sp., *Micrococcus* sp., *Flavobacterium* spp., *Bacillus* sp., and *Actinomyces*

TABLE 8.3
Predominant HPC Bacteria in Distribution System Biofilms

Bacteria	Occurrence in				
	Water Column	Zinc Floc Sediment	Flushed Sediment	Ion Tubercle	Pipe Surface Scraping
Pseudomonas vescularis	++	++	+		++
Flavobacterium spp.	++	++	+		++
Pseudomonas diminuta	+			+	
Pseudomonas cepacia					+
Pseudomonas pickettii					+
Pseudomonas stutzeri			+		+
Pseudomonas fluorescens			+	+	
Pseudomonas putida				+	
Pseudomonas paucimobilis	+			+	
Pseudomonas maltophila	+				
Alcaligenes spp.					+
Acinetobacter spp.	+				+
Moraxella spp.	+		+		+
Agrobacterium radiobacter					+
Arthrobacter spp.	+				+
Corynebacterium spp.			++	++	+
Bacillus spp.				+	+
Yeasts					+
CDC group 11 J	+				
Enterobacter agglomerans					+
Micrococcus spp.					

Legend: + = Positively identified in low numbers; ++ = positively identified in high numbers.

Source: Adapted from Geldreich.[116]

sp. as well as some yeasts.[41] *Flavobacterium* are known to produce bacteriocin substances which are well known to be inhibitory to coliform bacteria.[42] It has also been observed that in chlorinated waters which contain high numbers of antagonists, low numbers of coliform can be isolated.[43] Conversely, when the HPC (antagonist) population is low (less than 20% of the total plate count), a greater coliform incidence (75%) is recorded. It is thought that such suppression can result in an underestimation of coliform numbers by as much as 80% in potable water.[41,44]

The effects of heterotrophic bacteria on the public health is not known.[45] However, it is well recognised that high heterotrophic plate counts detected in home filtering devices are capable of generating gastrointestinal illness.[31] Also, research by Payment, Franco, and Siemiatycki in 1993[46] found that despite drinking water that had met all bacteriological and physcio-chemical quality standards, a cohort of individuals which were studied still showed significant levels of gastro illness. From this work, it was established that there was a weak association between heterotrophic bacteria and these illnesses with bacteria growing at 35°C associated with this effect.

TABLE 8.4
Reported Occurrence of Pigmented Bacteria in Potable Water

Water source	Bacterial Count (Cfu per ml)	Pigmented (%)
Bottled water	140–570,000	0–100
Distribution water	1,000	80–90
Distribution water (Cl_2)	200	62
Distribution, well water	500	35
Reservoir water (Cl_2)	5–150,000	55–90
Well water	30–690	10–14

Source; Adapted from Geldreich.[116]

This and other sporadic cases of research raises the question of bacterial numbers within water distribution systems and their association with gastrointestinal illnesses. More fundamental, however, is the fact that HPC in excess of 500 cfu per ml has been shown to mask the presence of coliforms within potable water.

Opportunistic pathogens which form part of the HPC and are associated with potable water include *Legionella* sp., *Aeromonas* spp., *Flavobacterium* sp., and *Pseudomonas aeruginosa*. Also some species of *Klebsiella* and *Serratia* have been included in this group.[47,48]

Pigmented bacteria constitute at least 80 to 100% of the HPC bacteria. Reported occurrence of pigmented bacteria evident in potable water is shown in Table 8.4. The health implications of these pigmented bacteria found within potable water has been expressed over recent years as an indicator of poor treatment practice, particularly because of their ability to withstand the disinfection processes employed within potable water. Pigmented bacteria which have been isolated from potable water are numerous and include: *Pseudomonas*, *Flavobacterium*, *Xanthomonas*, *Lysobacter*, *Flexibacter*, *Mycobacterium*, *Corynebacterium*, *Cytophaga*, *Empedobacter*, *Nocardia*, *Erwinia*, *Enterobacter*, *Serratia*, *Micrococcus*, and *Arthrobacter* sp. [49,50] Of the genera mentioned, *Flexibacter*, *Cytophaga*, *Lysobacter*, and *Arthrobacter* have not been documented as causing any form of human disease.

The pigmented bacteria most predominately isolated from potable water include: *Pseudomonas*, *Flavobacterium*, *Xanthomonas*, *Corynebacterium*, *Nocardia*, *Enterobacter*, and *Staphylococcus aureus*. These bacteria will now be discussed in greater detail.

8.7.2 *Pseudomonas*

Pseudomonads are aerobic, gram-negative bacilli, and nonspore forming. They are oxidase and catalase positive and motile by use of polar flagella. Of the 100 species presently recognised, the most important medical strain is that of *Pseudomonas aeruginosa* which generally produces a distinctive green pigment (pyocyanin) on agar, but others, namely, yellow, blue, red, and black, have been cited. *Pseudomonas aeruginosa* is able to grow at 41°C and produces O and H antigens which are used for serogrouping. Particular groups of *Pseudomonas aeruginosa* known to be pathogenic include serogroups 11 and, possibly, 9. They have been isolated and subsequently identified in trauma patients recovering from burns, intensive surgery, and cancer therapy.[51-54] Surveys undertaken within potable water supplies have shown this organism to constitute around 2 to 3% of the total HPC, and it is responsible for some 10% of nosocomial infections. It can be recovered in low numbers within potable water.[55]

Other *Pseudomonas* species often isolated from potable water include *fluorescens, alcaligenes, mendocina, putida, cepacia, allei, maltophila, testosteroni, vescularis, flava, pseudoflava, palleroni, rhodos, echinoides, radiora,* and *mesophilica.*[45,56,57] Also, *Pseudomonas fluorescens,* which is an important contaminant of blood supplies, has been known to cause septicemia occasionally.

Outbreaks owing to *Pseudomonas aeruginosa* isolated from potable water have been identified.[58-60]

8.7.3 *Flavobacterium*

Flavobacterium spp. are gram-negative oxidase-positive, aerobic, nonfermentative organisms.[61] Many of the species form bright yellow colonies and others have been shown to form red or green colonies on normal HPC agar.[62-64]

It is a genera which inhabits food, water, and soil and also has been isolated from the normal resident flora of the oropharynx. The most important clinical strain is *Flavobacterium meningosepticum* which causes a number of effects in humans including septicemia, wound infections, endocarditis, and respiratory tract infections together with meningitis in young children. They have been identified in potable water supplies, nebulizers, humidifiers, hemodialysis equipment, and drinking water fountains.[62,65,66]

Conditions that favour the growth of this organism in potable water includes the absence of free chlorine, water temperatures above 15°C, and often static water conditions. Many *Flavobacterium* spp. have been shown to be both chlorine and copper resistant with some species known to cause abdominal cramps after drinking from a water fountain.[67]

8.7.4 *Xanthomonas*

One of the most predominately cited organisms isolated from potable water is *Xanthomonas maltophila.* It has also been isolated from clinical samples and has been known to cause septicemia, pneumonia, wound infections, and many other diseases.

8.7.5 *Corynebacterium*

Corynebacterium is associated with the normal resident flora of human skin and the upper respiratory tract. They are regularly isolated from potable water and biofilms.[68-70] The most important clinical strain is that of *C. diptheriae* which does not produce a normal pigmentation on standard HPC agar but does produce a grey/black pigment on tellurite blood agar.

8.7.6 *Nocardia*

Nocardia is a common isolate of soil and water and also has been isolated from the air. It generally produces a yellow-orange pigment on HPC agar. They are not known to be primary pathogens but are generally now being considered as emerging opportunistic pathogens associated with patients undergoing organ transplants, steroid users, AIDS patients, and people with diabetes.

8.7.7 *Enterobacter*

The two most frequently isolated species from humans include *Enterobacter cloacae* and *E. aerogenes*. *E. cloacae* has been found in potable water systems and is often found attached to particles.[71] They are usually pigmented yellow and known to cause neonatal meningitis and sepsis, though they are low on the list of common causes of meningitis and septicemia.

8.7.8 *Staphylococcus aureus*

Staphylococcus aureus is a gram-positive bacteria which is a normal component of human skin and feacal waste. The health implications of *S. aureus* generally lie in individuals who are exposed to it for long periods of time such as in whirlpools and dental therapy. Of concern is the thermostable enterotoxin particularly associated with food preparations. Some studies on food poisoning incidents have been linked to water supply contamination.

8.7.9 Treatment of Infection

Most cases of waterborne diseases owing to the heterotrophic bacteria involve sensitivity tests to find the most appropriate antibiotic for treatment. However, most infections just require supportive treatment and fluid replacement when necessary. For *Pseudomonas aeruginosa*, however, these are generally resistant to antimicrobial agents and, therefore, therapy should be based on antibiotic sensitivity test results. Ciprofloxacin is showing promise in being effective against *Pseudomonas aeruginosa* in severe cases of infection.

8.7.10 Occurrence in Biofilms

A large number of pigmented heterotrophic bacteria have been associated with biofilms in potable water. With the one exception being *Pseudomonas aeruginosa*, these organisms are generally infrequent causes of disease in humans when compared to the effects of primary pathogens.

The colonisation of private water supplies are the most frequent supplies of *Staphylococcus aureus* infections, particularly supply lines and water storage tanks.[43] *Flavobacterium* spp. are well known to colonise drinking water pipes and are often found to be resistant to the normal chlorination process. The only public health incidence associated with *Flavobacterium* and biofilm growth was reported in 1966.[67] Investigation from this study revealed that colonisation of *Flavobacterium* sp. had taken place from copper pipe systems which led to a drinking water fountain system.

Another important public health problem also being identified in potable water is the fact that a number of heterotrophic bacteria are often found to be antibiotic resistant. It is possible that these bacteria may present a public health problem if genetic resistance is transferred to possible pathogens. The transfer of resistance factors has been identified in source water.[72,73] Whilst water treatment processes may remove these organisms, it is well documented that this resistance factor can be detected in a number of genera ranging from *Moraxella*, *Acinetobacter*, *Staphylococcus*, and *Pseudomonas* spp.[74]

8.7.11 Total and Faecal Coliforms

8.7.11.1 General Characteristics

Total coliforms are present at high levels within water contaminated by animal and/or human feaces. Coliform species identified in various potable water supplies is shown in Table 8.5. As a group, coliform bacteria do not cause disease[47] but are usually present when enteric pathogens are evident. This is the reason why coliforms are used as a primary indicator of potable water quality. As a group, they are used to determine the efficiency of water treatment and integrity of potable water, and they are used as a screen for fecal contamination.

Coliform or enteric bacteria consist of a heterogeneous group of gram-negative aerobic bacilli which are commensals of the gastrointestinal tract of mammals. The family includes *Escherichia*, *Citrobacter*, *Enterobacter*, *Klebsiella*, *Proteus*, *Serratia*, *Shigella*, and *Salmonella*. The predominant species are generally *Enterobacter cloacae*, *Klebsiella* spp., *Citrobacter freundii*, and *Enterobacter Agglomerans*.[75] The group are aerobic or faculatively anaerobic which optimally grow at 37°C. They are motile, bile tolerant, and readily grow on MacConkey's agar. On this agar, *Escherichia*, *Enterobacter*, *Klebsiella*, and *Serratia* produce pink colonies and *Salmonella*, *Shigella*, and *Proteus*, which do not ferment lactose, form pale colonies.

The coliform concept is considered to be the most reliable indicator used to show faecal contamination of potable water.[76] For many years, it has been questioned as to the use of coliforms as a measure of water quality owing to the fact that the coliform ideal was developed initially on decisions and assumptions which were largely correct in light of knowledge available at the time.[77] This, therefore, seems

TABLE 8.5
Coliform Species Identified in Various Public Water Distribution Systems

Citrobacter	*Enterobacter*	Escherichia	*Klebsiella*
C. freindii	*E. aerogenes*	*E. coli*	*K. pneumoniae*
C. diversus	*E. agglomerans*		*K. rhinoscleromatis*
	E. cloacae		*K. oxytoca*
			K. ozaenae

Source: Adapted from Geldreich.[116]

to suggest that such particular practices are no longer relevant to present day public health solutions.

Several deficiencies of the coliform group as indicator organisms have become evident, particularly owing to[8]

- Regrowth in distribution systems.
- Suppression by high background bacterial growth.
- Not being indicative of a health threat.
- A lack of correlation between coliforms and pathogen numbers.
- No relationship between protozoan and viral numbers.
- The occurrence of false positive and false negative results.

The most predominant fecal coliform is that of *Escherichia coli*, a bacteria which does not survive long in aquatic environments so their presence is used to indicate that there is fresh faecal contamination present.

8.7.11.2 *Escherichia coli*

8.7.11.2.1 General characteristics

Escherichia coli are motile, gram-negative, nonspore-forming bacilli which produce both acid and gas from lactose at 37 and 44°C. It is used to indicate contamination of potable water with either human or animal excreta. They ferment lactose or mannitol at 44°C with the production of gas after 24 hours.[78] A number of serotypes are evident; many of which are harmless species. Some less commonly encountered *E. coli* strains found both within the environment and potable water systems are capable of giving rise to diseases usually in the form of diarrhoea. The process by which *E. coli* causes diarrhoea varies between strains. These strains can be grouped depending upon which mechanism is used by a particular strain. These include the following.

Enterotoxigenic *E. coli* (ETEC) causes gastroenteritis with profuse watery diarrhoea accompanied with abdominal cramps and vomiting. They produce two enterotoxins, a heat stable toxin and a heat labile oligopeptide toxin. Both toxins enter the cell and increase the concentration of cyclic guanine monophosphate (cGMP) or cyclic adenine monophosphate (cAMP). This affects electrolyte transport leading to excessive fluid loss. The incubation period associated with infection of this group of *E. coli* is 12 to 72 hours, with duration of illness often 3 to 5 days.

It has been estimated that approximately 2 to 8% of the *E. coli* found in water are enteropathogenic *E. coli* which causes travellers diarrhoea (watery to profuse watery diarrhoea). Whilst water and also foods are instrumental in the transmission and spread of *E. coli*, the required dose for this pathogen to cause infection is high, typically in the range of 10^6 to 10^9 organisms. A biofilm is, therefore, a very good area for multiplication of this organism. The incubation period when infected with this group is 1 to 6 days with the duration of infection lasting 1 to 3 weeks.

Enterohaemorrhagic *E. coli* (EHEC) produces a shiga-like toxin that is cytotoxic to vero cells. The commonest EHEC is 0157. Following ingestion of the required dose, symptoms include watery and bloody diarrhoea associated with vomiting. EHEC can also cause two distinct conditions, haemorrhagic colitis and haemolytic uraemic syndrome. Strains produce a toxin which results in the development of bloody diarrhoea or acts systemically resulting in haemolysis and renal failure. The incubation period following ingestion is 3 to 8 days and duration of infection is 1 to 12 days.

Enteroinvasive *E. coli* binds to enterocysts of the large intestine where they destroy cells resulting in tissue damage. The incubation period is usually 1 to 3 days and the duration of infection is 1 to 2 weeks.

Other *E. coli* groups known to be infectious include Enteroaggregative *E. coli* (EaggEC) and Diffuse Adherence *E. coli* (DAEC) which adhere to Hep-2 cells aided by the fimbriae.

The incubation period associated with *E. coli*, if infected, depends on the type of strain one has been infected with. This is particularly owing to the different pathogenic mechanisms evident. Generally, the incubation period is 1 to 2 days, but it has been known to be up to 5 days. Most species cause dehydrating diarrhoea and traveller's diarrhoea.

8.7.11.2.2 Treatment of infection

In most cases of *E. coli* diarrhoea, the treatment required is generally supportive. However, in severe or prolonged illness, ciprofloxacin may be given.

8.7.11.2.3 Presence and outbreaks associated with potable water

Of grave concern to public health are waterborne outbreaks associated with *E. coli* 0157:H7. Whilst outbreaks are rare, a number have been reported[80] in Scotland[81] and in Cabool, MO, in 1990. These were owing to disturbances in the water distribution network. The outbreak which occurred in the U.S. resulted in 243 cases of diarrhoea and 4 deaths among elderly people.[82,83]

EHEC is well suited to waterborne spread because the infectious dose appears to be low, probably less than 100 organisms.

8.7.11.2.4 Occurrence in biofilms

Within the U.K., the number of coliform failures in water samples taken at customers' taps has decreased tenfold in the last 5 years. However, low numbers are still being detected in water supplies and continue to be a problem.[84] This probable source is regrowth of coliforms within biofilms,[85,86] causing coliform failures as a result of continual shedding of bacteria into the bulk water.[87]

LeChevallier[40] has identified several factors that show coliform growth in distribution networks as a result of biofilm growth. These are

- No coliform detected in treatment plant effluent indicating that the treatment plant is not the source of coliform bacteria.
- High coliform densities are detected in a potable water system despite the fact that very low numbers were entering the system from the treatment plant. A study by LeChevallier et al.[43] on biofilms found that even effluent contained low coliform levels (0.3 cfu/100 ml) and that coliform densities increased twentyfold (0.64 cfu/100 ml) as the water moved from the treatment plant through the distribution system.
- The reoccurrence of coliforms lasted for a prolonged length of time. This can be largely attributed to the difficulty of inactivating biofilms.

The phenomenon of coliform regrowth is a significant problem for water companies because coliform bacteria may mask the presence of indicator organisms, resulting in the breakdown and effectivity of adequate treatment provision.

Many researchers have studied the fate of coliforms in experimental potable water systems with conflicting results.[86,88-90] However, the consensus seems to be that coliforms are able to survive and grow in biofilms.[91]

Several investigators have shown chlorine levels used in water treatment are inadequate to inactivate biofilms which have incorporated coliforms. It is clear that routine chlorination practices, which rely on a residual effect, will not prevent or control bacterial regrowth. It has been suggested that monochloramines might be more effective for biofilm control than chlorine. It has a longer lasting disinfecting effect than chlorine but is, however, less effective than free chlorine.[92] It has been found that particular bacteria, such as *E. coli* are able to withstand 2400 times more chlorine when attached to a surface than as free cells in water leading to high survival rates in potable water.[93,94] In contrast, Sibille[90] has found that grazing portozoan may have been responsible for the removal of *E. coli* from a potable water system which is more rapid than when the protozoan are absent.

As previously cited, potable water biofilms may serve not only to allow the growth of potentially pathogenic microorganisms[95] but also in the development of a functional niche protecting coliforms from harsh or aggressive environments.[96] One study has recently suggested that contrary to the widely accepted understanding that coliforms survive and grow in biofilms on pipe walls, this is not true.[91] Overall, however, of the studies undertaken, it has been found that coliforms are capable of growing in biofilms despite the evidence of excessive chlorine levels present in the system.[13,32,97-102]

8.7.11.3 *Yersinia* spp.

8.7.11.3.1 General characteristics

Yersinia spp. have been implicated in a number of waterborne outbreaks and, therefore, would constitute incorporation into this chapter. They are gram-negative, facultative, nonspore forming bacilli and members of the Enterobacteriaceae group. As a group, they are nonmotile at 37°C, but motile at 22°C. Presently, there are 11 recognised species which have been identified, of which 4 are of human significance. These include *Y. pestis*, *Y. pseudontuberculosis*, *Y. enterocolitica*, and *Y. ruckeri*.

The most predominate species associated with potable water is *Yersinia enterocolitica* of which serotype 0:3 is the most infectious. Serotypes 0:3, 0:4, 32, 0:5, 27, 0:6, 30, 0:6, 31, 0:8, 0:9, and 0:21 are the pathogenic strains which have been linked to diarrhoea, often mimicking appendicitis.[103]

Infection doses of these bacteria are high at 10^9 organisms. The incubation period associated with infection with *Yersinia* is 2 to 7 days with the duration of infection between 1 and 21 days (with an average of 9 days).

8.7.11.3.2 Treatment of infection

For treatment of infections owing to *Yersinia*, it is generally found that specific antibody therapy can be given, but the mortality with septicaemia can be as high as

75% of the cases. Tetracycline is often recommended as the preferred antibiotic. However, a good response has been seen with ciprofloxacin.

8.7.11.3.3 Presence and outbreaks associated with potable water

Yersinia have been isolated from potable water supplies during a number of outbreaks. These have been identified in Germany,[104-106] suggesting a waterborne route of spread by this organism.

8.7.11.3.4 Occurrence in biofilms

To date there is no available information on the occurrence of *Yersinia* in biofilms.

8.7.11.4 ***Mycobacterium* sp.**

8.7.11.4.1 General characteristics

Mycobacterium are obligate aerobes or microaerophilic, noncapsulated, curved bacilli that are positively identified with the Ziehl-Neelsen acid-fast method. *Mycobacterium* are free living environmental saprophytes of which there are approximately 60 species known. The most recognised pathogens include *M. tuberculosis* and *M. leprae*. *M. avium, M. kanasii, M. fortuitum*, and *M. chelonae* which have been described as potential human pathogens with evidence documented that they are transmitted in potable water. *Mycobacteria* sp. are now becoming recognised as being opportunistic pathogens of significance,[107,108] particularly abundent in patients who are immuno-compromised.

Mycobacterium sp. cause a number of diseases in humans which range from soft tissue infections, cervical lymphadenopathy, septicemia, and pulmonary diseases.[109] The pathogenecity of this organism depends particularly upon the infecting agent and also the site of infection.

8.7.11.4.2 Treatment of infection

There are a number of antituberculous drugs available which are very effective against some waterborne mycobacterial species. However, the treatment may be as long as 2 years.

8.7.11.4.3 Presence and outbreaks associated with potable water

Mycobacterium sp. are well established in all types of aquatic systems ranging from groundwater to potable water and raw water.[107,109-111] As with a lot of outbreaks of waterborne illnesses, they are usually prevalent during treatment deficiencies. However, findings within potable water have revealed that these organisms have been resistant to the normal chlorination processes.[112] Experiments have shown that chlorine levels of less than 1.0 mg per litre may not, in fact, inactivate a number of these opportunistic pathogens.[113] These include *M. fortitum, M. gordonae, M. avium, Mchelonae, M. kansasii*, and *M. intracellulare*. Even a free chlorine residual of greater than 0.2 mg per litre at a pH of 5.9 to 7.1 had no effect at reducing acid-fast bacteria in potable water. Research has also found that *Mycobacterium* sp. are more resistant to ozone[114] and chloramines[115] than *E. coli*.

Studies have shown that *Mycobacterium avium* complex organisms (i.e., *M. avium* and *M. intracellulare*) can be isolated from potable water. These organisms are opportunistic and pose significant health implications to immuno-compromised patients. Mycobacterial infections, therefore, present the greatest risk to patients within hospitals, particularly the elderly, owing to aerosoled water.[108]

8.7.11.4.4 Occurrence in biofilms

Evidence of the growth of mycobacterium in biofilms, owing to the persistance of certain strains of *Mycobacteria* in potable water, have found this to be as long as 41 months, particularly during summer months.[116] Growth in biofilms has also been reported in old portions of piping which experience corrosion problems and elevated pH.[117] Some regrowth has also been reported in dead-ends, evident in water distribution systems where low chlorine levels are evident and total organic carbon and turbidity are high.[116] *Mycobacterium* are also very prominent in buildings' plumbing systems where water flow can be slow.[117] The presence of these organisms are also found on a continous basis within cold and hot water taps.[118]

To date three main studies have shown *Mycobacteria* can occur in biofilms,[119–121] indicating public health significance.

8.7.11.5 *Salmonella* sp.

8.7.11.5.1 General characteristics

Salmonella are members of the Enterobacteriaceae and regarded as one of the most important human pathogens in Western Europe. They are gram-negative, facultative, anaerobic bacilli and nonspore forming. Most are motile. Generally, they are differentiated biochemically into four subgenera (I to IV). Subgroup I are general human pathogens with the most serious diseases caused by *Salmonella typhi* (typhoid), a sergroup 9 organism, and *Salmonella paratyphi* (paratyphoid), serogroups 2, 4, or 7. Infection with *Salmonella typhi* is associated with fever, malaise, headache, coughing, nausea, vomiting, and abdominal pain. The incubation period ranges from 7 to 28 days but does vary substatially from outbreak to outbreak. The duration of infection ranges from weeks to months.

Salmonella gastroenteritis is associated with watery and occasionally bloody diarrhea (30% of cases). In some cases, nausea and vomiting may be present. If infection is severe, it can lead to death. The incubation period for *salmonella* infections is usually 8 to 48 hours, but it has been known to be as long as 7 days with symptoms lasting 3 to 5 days.

8.7.11.5.2 Treatment of infection

The treatment of *salmonella* gastroenteritis is generally supportive. Antibiotics are rarely given when infection occurs. However, in severe cases and prolonged disease, ciprofloxacin may be given.

For typhoid, the drugs often prescribed are ciprofloxacin and chloramphenicol. Others may be given including co-trimoxazole or amoxycillin.

8.7.11.5.3 Presence and outbreaks associated with potable water

Salmonella are virulent pathogens affecting public health and are present predominately in raw water and occasionally in potable water. Chlorine has been highly effective at destroying this organism in the planktonic state. However, in potable water some *Salmonella* sp. have been found to be very resistant to certain types of disinfectants.[122]

8.7.11.5.4 Occurence in biofilms

In terms of biofilms, Jones and Bradshaw[123] have studied *Salmonella enteritidis* in systems modelling water distribution systems, and they have demonstrated that it formed dense metabolically active biofilms. In the food industry, Hood and Zottola[124] have shown developing biofilm on stainless steel. Korber et al.[125] investigated *S. enteritidis* and have found that surface roughness and the timing of disinfectant application were important factors controlling the presence of the biofilm.

8.7.11.6 *Shigella* spp.

8.7.11.6.1 General characteristics

Shigella is a major cause of bacterial dysentery (shigellosis) within the U.S.,[126] with cases now stabilising within the U.K. They are gram-negative, nonmotile rods, oxidase-negative, and members of the Enterobacteriaceae. Despite the group consisting of over 40 species, only *Sh. dysenteriae*, *Sh. sonnei*, *Sh. flexneri*, and *Sh. boydii* are able to cause gastrointestinal disease. In most cases, the incubation period is 1 to 7 days with the disease symptoms being dependant upon the strain. In general, infection is associated with dysentery and fever which may last 4 to 7 days. With *S. sonnei*, infection has no symptoms and those that develop diarrhoea usually settle in a few days. In severe cases, diarrhoea can be associated with vomiting. Other groups of *Shigella* spp. produce severe diarrhoea which can persist up to a week with the stools often covered in mucus and blood. Severe diarrhoea is often owing to *S. dysenteriae* type 1.

8.7.11.6.2 Treatment of infection

In minor cases of infection with shigella, the infection is self-limiting and requires no form of therapy other than oral fluids. Antibiotics are rarely given but may be administered in seriously ill individuals.

8.7.11.6.3 Presence and outbreaks associated with potable water

The epidemiology of *Shigella* is similar to that of *Salmonella* except that it rarely infects animals and does not show the same resilience as *Salmonella* in the environment. The mode of spread seems to be person-to-person, often characterised as a disease associated with overcrowding. The incidences of shigellosis have been increasing, particularly as a result of poor quality drinking water which has become contaminated by sewage.[127] A number of cases has been linked to potable water. These include a Caribbean cruiseliner,[128] public school,[129] and domestic household in the U.S.[130]

8.7.11.6.4 *Occurence in biofilms*

As yet there is no available literature of the association with biofilms with potable water.

8.7.11.7 ***Campylobacter*** **spp.**

8.7.11.7.1 *General characteristics*

Campylobacter spp. was first discovered in the late 1970s as a human disease-causing agent[131] and it has been relatively recent that the bacteria has been cultured.[132] Considering this, it is now the most common bacterial cause of gastroeneteritis. The disease caused by *Campylobacter* is called Campylobacteriosis. It is associated with abdominal pains, fever, and vomiting[133] and has an incubation period of 3 to 5 days. The duration of illness normally lasts from 1 to 4 days but has been known to last over 10 days. This would possibly signify why the epidemiology of *Campylobacter* with respect to the human host is not fully elucidated.

Campylobacter is gram-negative, curve-shaped bacteria, microaerophilic, motile, and 2 to 5 μm in length. It is oxidase positive, nonsporing, and a nitrate reducer but does not produce acid in the presence of carbohydrates and requires microaerophilic atmosphere plus 10% CO_2 for growth. There are two main subgroups. Subgroup 1 consists of *C. coli*, *C. jejuni*, *C. laridis*, and *C. upsaliensis*. Subgroup 2 consists of *C. fetus*, *C. hypointestinalis*, *C. concisus*, *C. mucosalis*, and *C. sputorum*. Subgroup 1 are thermophilc *Campylobacter* and are able to grow at 43°C. Of this group *C. jejuni* is the commonest cause of gastroenteritis. The incubation period associated with *C. jejuni* is 2 to 4 days. Cross-contamination from animals to humans is a very important mode of transport, particularly as there is a large reservoir of *Campylobacter* known to be evident in cattle, poultry, wild birds, and domestic pets.

A total of 40,161 reports of *Campylobacter* infections in England and Wales were received by the Communicable Disease Surveillence Centre (CDSC) in 1996 compared with 50,201 received in 1997. This 20% increase suggests that *Campylobacter* would be a likely candidate for establishing itself in potable water environments. Within the U.S. it was first reported in 1978 and as a result 200 people where infected. During 1987 in the U.K., there were 27,000 reported cases of *Campylobacter enteritis.* This figure, however, rose to 30,000 in 1990. It is now becoming established that *Campylobacter* spp. is the major source of gastroenteritis in Europe, the U.S., and other parts of the world.[34-136] Within the U.S. it is estimated that the incidence of this organism is 30 to 60% per 100,000 of the population.[137]

8.7.11.7.2 *Treatment of infection*

Treatment of *Campylobacter* infection generally requires fluid and electrolyte replacement. However, ciprofloxacin and erythromycin have been used for treatment. Generally, most cases recover without treatment.

8.7.11.7.3 *Presence and outbreaks associated with potable water*

Campylobacter is well known to be a microbial coloniser of surface waters[138] with recovery from these sites of origin being very high, particularly during autumn and winter, suggesting this organism is affected by seasonal changes.[138]

Whilst concern is expressed with respect to untreated or poorly chlorinated water, isolation of this organism from properly treated and disinfected water has not been reported. Such information suggests that conventional water treatment processes may be sufficient for preventing *Camplylobacter* spp. development in water supplies.[135]

8.7.11.7.4 Occurence in biofilms

Work by Buswell et al.[139] has demonstrated that although *Campylobacter* cells are environmentally sensitive, their survival in a culturable form in water biofilms can be sufficiently prolonged for this vehicle to be potentially important in the transmission of this pathogen. In the poultry industry, the presence of *Campylobacter* is a problem. Results from 15 randomly chosen poultry flocks in commercial processing plants showed that all flocks were contaminated initially with thermophilic *Campylobacter* spp.[140] When comparing flocks before and after chlorinated-water sprays were used to limit microbial contamination on equipment and working surfaces, it was found that numbers of campylobacter in biofilms on packaged carcasses were significantly lower after the changes had been made. The flocks themselves may get contaminated by *Campylobacter* spp. from the environment with the drinking water being a potential route.

8.7.11.8 *Legionella* spp.

8.7.11.8.1 General characteristics

Legionella was first described in 1976 when 182 members of a Pennsylvania American Legion convention were infected. *Legionella* are bacteria of which 40 different species have been identified. They are short, but sometimes filamentous, gram-negative, aerobic bacilli which grow optimally at 35°C. Legionellosis, the disease caused by *Legionella*, is recognised worldwide and, recently, there have been several large outbreaks in the U.K. Legionnaires' disease is a severe form of pneumonia with a low attack rate but with a mortality rate of about 10%. *Legionella* infection can also be associated with Pontiac fever which is a mild nonpneumonic flu-like illness with a high attack rate. Whilst this disease is recognised as an acute pneumonia with a high fatality rate, it is known also to affect the nervous system, gastrointestinal system, and urinary system. The incubation period for Legionnaires' disease is usually between 2 to 14 days with the infection lasting weeks to months. The symptons of infection include pneumonia with anorexia, malaise, myaglia, headache, rapid fever, chills, coughing, chest pain, abdominal pain, and diarrhoea. With Pontiac fever, symptoms include fever, chills, myalgia, and headaches. The incubation period for this disease is 5 to 66 hours and can last from 2 to 7 days.

Legionella pneumophila causes the majority of infections reported in the U.K. and is found commonly in both natural and man-made water systems. This species consists of 14 serotypes with a large difference in virulence to man between them. Standards for the control of *Legionella* in water systems are given in an Approved Code of Practice, 1991, published by the Health and Safety Commission.

Research within potable water systems has been unable to establish levels or concentrations of *Legionellae* which could instigate a disease. Attributable risk owing to exposure within the domestic environment is needed together with a greater understanding of the ecological factors which contribute to its virulence. Being tolerant to high temperatures, this organism becomes suited to temperatures associated with man-made water systems.[141] An extensive study on *Legionella* within domestic environments was conducted in Germany.[142] This established that within the home it is important to consider the entire home plumbing system and not just the hot water system when evaluating the risk of *Legionellae*.

The organism can be transmitted to an unsuspecting host generally via aerosolisation and ingestion. Aerosolisation is particularly relevant in cooling towers, humidifiers, and showerheads with ingestion relevant when it is evident within potable water.[143] However, as yet, there is no available information of consumption in potable water being attributed to the disease. Of particular concern is that of nosocomial infection of potable water supplies within hospitals, particularly when it is known to become established within a biofilm.[144]

Research has found that *Legionella pneumophilia* is capable of growth within host organisms called protozoa, commonly found in water systems. This undoubtedly offers a protective niche to *Legionella* against adverse environmental conditions, and research has shown that when *Legionella* emerges from its hosts, it appears to be much more resistant to the action of biocides than bacteria growing under natural laboratory conditions. This has implications for the measures used to control and eradicate *Legionella* in potable water systems. Generally, free *Legionella* is destroyed easily by ultraviolet light and biocides. In environments where nutrient levels are low, recent research has shown that considering the presence of a conditioning film which has a higher nutrient content than the bulk liquid phase, the association of *Legionella* with a biofilm is highly probable. Therefore, the degree of colonisation of a system by *Legionella* should include the biofilm.

8.7.11.8.2 Treatment of infection

Treatment is usually started on the basis of clinical suspicion. The antibiotic most often given is erythromycin. However, ciprofloxacin or rifampicin may be added. With Pontiac fever, most patients recover without specific therapy.

8.7.11.8.3 Presence and outbreaks associated with potable water

Whilst *Legionella* has been isolated in domestic households, it is worth noting that consumers' water supplies despite its presence there are often no apparent signs of disease.[145] This presence could be a very important health effect, particularly if numbers develop to high concentrations within biofilms and, thus, are sloughed off at the consumers' taps. Control in large municipal buildings and domestic water systems involves cleaning pipes on a regular basis. Hot water systems should be stored preferably at 60°C with a minimum temperature of 50°C.[141] However, scalding is a public health effect induced by these high temperatures. Detailed procedures for precautionary measures are available for both hot water and hot water supplies.[146]

Whilst it is practical for baselines of disease to be obtained in the home for establishment of working methods with larger premises, this should be done as part of a management regime.[147] Modern buildings with extensive water distribution and air conditioning systems where water temperatures of 30 to 40°C are encountered, have created ideal breeding grounds for *Legionella*. Such systems have the potential for disseminating *Legionella* through aerosol formation to a large number of people as has been highlighted by recent outbreaks of Legionnaires' disease.[148,149]

One source of protection for *Legionella* in potable water systems is within amoeba. If conditions become unfavourable for amoeba, that is, food availability is poor or the amoeba are subjected to dry conditions, the amoeba forms a hard and impervious outer protective shell called a cyst. If *Legionella* is found within this cyst, it has excellent protective environments against dry conditions, extremes of temperature, and treatment with biocides.[150] As the bacterium is within the cyst, it can be blown away in the air. It has been established that the cyst is able to survive for nearly 400 days in sterile tap water without additional nutrients.[151] If conditions become variable for growth, the cyst is able to break open and amoebal cells together with *Legionella* are released. This, therefore, provides a perfect protective environment within very harsh conditions, suggesting good hygienic practices are essential. Water temperature and also plumbing material have been an important factor for the growth of both the protozoa and *Legionella*, itself.[157] It is well established that *Legionella* and amoeba grow best at 30 and 40°C. It is found that when thermostats on hot water tanks in homes have been set to below 55°C, amplification of this opportunistic pathogen is able to occur, often being released in aerosols which can be inhaled easily. The usage of recommended temperatures within hot and cold water supplies in large buildings has helped to reduce the epidemiological spread of *Legionella*.[153] It is also recommended that cold water should be stored and distributed below 20°C because *Legionella* is unable to multiply below this temperature. Within hot water systems, *Legionella* is only able to survive for a few minutes above 50°C, unless, as mentioned previously, it is found within a cyst.

It appears that *Legionella* grown inside amoeba is more able to survive biocide inactivation when compared to *Legionella* bacteria grown on artificial culture media.[154] It suggests that results of tests to assess the activity of biocides which are normally carried out on artificial culture media may not relate to the situation in the natural environment. Hyperchlorination at 10 mg per litre has shown to be effective at eradicating *Legionella* within potable water systems.[153] However, it is generally hard to establish and maintain these levels. Therefore, colonisation of *Legionella* on pipe surfaccs is feasible. If cysts are present containing *Legionella*, difficulties can arise. It has been shown that amoeba containing *Legionella* can withstand 50 mg per litre of free chlorine with viable *Legionellae* being recovered from these cysts.[150]

The eradication of *Legionella* is not a question of just adding biocide, it is now apparent that biocides must first kill the protozoa or else *Legionella* has the ability to establish itself in biofilms. Whilst the basic guidelines for removing *Legionella* in water systems are documented,[146] it is becoming apparent that action on biofilms may be more of a future investment.

For control of *Legionella*, generally, good management practice is required with respect to regular inspection, maintenance, and cleaning as well as the incorporation of biocides.

8.7.11.8.4 Occurrence in biofilms

Legionella present within a potable water system is likely to occur in biofilms, particularly where symbiotic relationships with other heterotrophes are evident. This is because bacteria such as *Flavobacterium breve*, *Pseudomonas*, *Alcaligenes*, and *Acinetobacter* are known to provide the nutritional requirements necessary for the long term survival of *Legionella*.[155]

Biofilms have been shown to be a contributory factor in supporting the growth of *Legionella* within potable water environments.[156] A better understanding of this amplification process may help to limit the multiplication of *Legionella* in water and, thus, interrupt the chain of events constituting its transmission.

Legionella is closely associated with a range of other microorganisms such as fungi, bacteria, algae, and protozoa. Since nutrient levels in water are usually low and the biofilm mode of growth has mutual benefits to microorganisms by providing a common food source, this provides an ideal solution to an otherwise stressful survival.

Studies have found that *L. pneumophila* can survive in biofilms for long intervals (20 to more than 40 days) at 24 and 36°C.[157] It has also been found that PVC is the best supporting material for *Legionella* in potable water.

8.7.11.9 *Aeromonas* spp.

8.7.11.9.1 General characteristics

Aeromonas spp. have been highlighted recently as to their role as primary agents for gastroenteritis. However, controversy still presides over their pathogenicity and epidemiology. This seems to be particularly associated with their taxonomy. Despite this, in light of the available literature it seems more likely that a number of the species within the genus, specifically *Aeromonas hydrophila*, deserves recognition as candidates involved in gastroenteritis and related diarroheal infections.

Aeromonas spp. are gram-negative, nonspore forming, facultatively anaerobic bacilli. They are generally classified as being nonmotile. However, there are three species of human significance which are thought to be motile, namely *Aeromonas hydrophila*, *Aeromonas caviae*, and *Aeromonas sobria*.

The role of these organisms and their association with diarrhoea remain very unclear. Despite this, a number of toxins have been isolated, namely, cytotoxic B-haemolysin, a cytotonic enterotoxin, an enteroinvasive factor, and a number of haemagglutins, proteases, and elastases. The mesophilic *Aeromonas* spp. are causative agents of human gastroenteritis, wound infections, septicaemia, soft tissue infections, and a number of other conditions.[158-161] The main clinical feature of *Aeromonas* includes diarrhoea, but this varies upon the specific species-causing infection. A number of infected individuals may also develop fever, bloody diarrhoea, and abdominal pain with some patients developing chronic colitis. Once infected, the incubation period associated with *Aeromonas hydrophila* is unknown, but once symptoms are evident, these may last for 42 days on average. Diagnosis of infection involves culturing the organism from faeces on ampicillin containing blood agar. The confirmation that the organism is present is made available with the use of biochemical tests.[162]

Ingestion of *Aeromonas* spp. may lead to diarrhoeal disease which is associated with the production of an enterotoxin.[163] Contrary to a number of findings, a study by Morgan et al., in 1985,[164] using human volunteers and five enteropathogenic strains of *Aeromonas hydrophila*, demonstrated diarrhoea in only 2 of the 57 persons infected with doses of 10^4 to 10^5.

8.7.11.9.2 Treatment of infection

Cases of infection with *Aeromonas hydrophila* have been treated with specific antibiotics which have often led to clinical improvements to the patient. The choice of antibiotic used in this way will be dependant upon the results of antibiotic sensitivity tests.

8.7.11.9.3 Presence and outbreaks associated with potable water

Aeromonas spp. are particularly prevalent within potable water and have been implicated as causative agents in waterborne outbreaks.[165-168] However, the possible role of potable water in the transmission of infections from *Aeromonas* is still under discussion.

Aeromonas hydrophila, within treated and untreated water, is the most frequently isolated phenospecies[169] and is present more commonly in untreated water than faecal coliforms.[170] Its significance in potable water is that for many years it has caused false reactions in total coliform counts. It is not an organism which is routinely investigated for within potable water despite evidence that this organism is an emerging problem in affecting water quality. Nevertheless, *Aeromonas* has been used as an indicator of the quality of water and because there is no correlation between its concentration and the level of faecal contamination, it has been suggested by many researchers that the quantification of *Aeromonas* spp. within potable water should be included as an indicator of potable water quality.[171-174] Reasoner[17] has also implicated this organism as a potentially waterborne pathogen for the future.

Surveys have been carried out on the prevalence of *Aeromonas hydrophila* in potable water. One study looked at 286 samples taken from taps and storage tanks in hospitals in the U.K. From this study it was concluded that the presence of *Aeromonas hydrophila* was recorded in 25% of the samples during summer but only 7% in winter.[175] A U.K. Water Industry Research Company contract with the Public Health Laboratory Service analysed the database relating to the reports of *Aeromonas* cases for 1989 to 1994.[176] The results revealed that from 1991 to 1994, 2718 isolates of *Aeromonas* sp. were recorded.

Aeromonas spp. within potable waters are best controlled by the use of chlorine levels at or above 0.2 mg per litre. However, recent studies have found that *Aeromonas* can be recovered when exposed with chlorine at levels of 0.5 mg per litre [177] and 2.5 mg per litre.[178]

8.7.11.9.4 Occurence in biofilms

Potable water is known to be a source of mesophilic *Aeromonas*, particularly, as a result of aftergrowth.[179-185] This aftergrowth has been shown to be a result of *Aeromonas* growth within biofilms.[10,186]

8.7.11.10 *Vibrio cholerae*

8.7.11.10.1 General characteristics

Vibrio are a group of bacteria which are gram-negative, comma-shaped, facultatively anaerobic, oxidase positive, and motile. They are exclusively transported and transmitted through water and able to grow at a pH of 8.6.

Vibrio cholerae have been divided into over 80 0-serovars although strain 01 servar is the most potent owing to the production and release of an enterotoxin which is a known cause of mild to profuse diarrhoea and vomiting resulting in excretion of large amounts of fluid in a short space of time.[187] Serotype 0139 is now widely distributed in water and threatens to become the eighth worldwide cholera epidemic.[188]

Vibrio cholerae 01 has an incubation period of 9 to 72 hours and produces profuse, watery diarrhoea and vomiting. These symptoms usually last for 3 to 4 days. With non-01 *Vibrio cholerae* the incubation period is longer, often 5 days. Clinical symptoms include watery diarrhoea which lasts 3 to 4 days.

8.7.11.10.2 Treatment of infection

The disease is relatively easy to treat. Generally, fluid and electrolyte treatment is given. The World Health Organisation recommends for treatment following infection, a solution of glucose and salts containing 3.5 g NaCl, 2.5 g $NaHCO_3$, 1.5 g KCl, and 20 g glucose in a litre of water. The use of antibiotics may shorten the duration of diarrhoea but are not given in mild cases.

8.7.11.10.3 Presence and outbreaks associated with potable water

This bacterium is now rare within both the U.K., Europe, and the U.S., owing to the use of water treatment procedures. However, it is still classed as an endemic in Africa and Asia. It is a naturally occurring bacteria found in sessile states on zooplankton and phytoplankton (e.g., Volvox).

Other species of the genus can also cause diarrhoea, although less severe than type 01. These include *Vibrio parahaemolyticus*, *Vibrio fluvialis*, and *Vibrio mimicus*, which are found in brakish and saline waters and are rapidly inactivated in high acidity waters.

8.7.11.10.4 Occurence in biofilms

No evidence of growth within a biofilm is available at present.

8.7.11.11 *Helicobacter pylori*

8.7.11.11.1 General characteristics

Helicobacter is a gram-negative, rod-shaped bacteria, motile, nonsporing, noncapsulated, and microaerophilic. They live on and between gastric epithelial cells below a layer of mucus. The organism has an ability to secrete urease which breaks down urea from the diet into ammonia and carbon dioxide. *Helicobacter pylori* was first cultured in 1982 and originally classified as *Campylobacter pyloridis* but is now assigned to the genus *Helicobacter*.[189] *Helicobacter pylori* is probably now the world's most common infection, with an estimated incidence of 50% worldwide.[190] The transmission routes and the reservoirs of *Helicobacter pylori* are unknown. However, it has been suggested that water may well be a possible reservoir of infection.[191]

Recent surveys have identified respiring *H. pylori* in surface water and groundwater, although the extent and epidemiological significance of these reports remain unclear. One major problem in the study of *H. pylori* has been the lack of a suitable direct method for the detection in water samples.

Symptoms associated with infection include nausea and abdomial pains which last between 3 and 14 days. Gastritis develops, which may be evident for a number of years.

8.7.11.11.2 Treatment of infection

At the moment, the method of treatment for *Helicobacter* infection is a combination of amoxycillin and metronidazole plus bismuth administered for about a month

8.7.11.11.3 Presence and outbreaks associated with potable water

Helicobacter pylori has been identified within drinking water.[192,193] They also have the ability to enter a viable but nonculturable state in many different environmental conditions[191] which makes conventional isolation and identification techniques inadequate to enable exact prevalence in drinking water. It is generally found that *H. pylori* is readily inactivated by free chlorine, suggesting that this organism can be controlled by disinfection practices normally employed in the treatment of potable water.[194] A recent study has shown *Helicobacter pylori* can survive for up to 10 days in milk at 4°C storage but only for 4 days in tap water.[195] This confirms that potable water may act as a vehicle for transmission of this organism.

Overall, the transmission of *Helicobacter pylori* within water is looking very likely, particularly so, within developing countries where contaminated drinking water and poor sanitation and hygiene conditions enable the organism to be transmitted by the faecal–oral routes. However, better methods for isolation and identification are necessary to fully substantiate this.

8.7.11.11.4 Occurrence in biofilms

Studies using mature heterotrophic mixed-species biofilms, using continous chemostat systems, have shown that when challenged with *H. pylori*, the bacteria can be associated with the biofilm for up to 192 hours post challenge.[196] This suggests that biofilms in potable water systems could be a possible and unrecognised reservoir of *H.pylori.*[197]

8.7.11.12 Fungi and Yeasts

8.7.11.12.1 General characteristics and association with potable water

The most common types of fungi found in potable water are filamentous fungi.[198-201] However, little regard has been given to their occurrence in potable water. From

TABLE 8.6
Filamentous Fungi and Yeast in Potable Water Supply

Microbial Genera or Species		
Fungi	**Yeast**	**Occurrence (%)[a]**
Cladosporium		19.4
Phoma		13.3
	Candida parasilosis	11.7
	Rhodotorula rubra	5.6
Alternaria		5.1
Exophiala		5.1
Aspergillus		4.6
	Aureobasidium pullulans	3.6
Penicillin		3.6
	Cryptococcus albidus var. *diffluens*	3.1
Cephalosporium		2.0
Drechslera		2.0
Phialophora		2.0
	Cryptococcus albidus var. *albidus*	1.5
	Cryptococcus uniguttulatus	1.5
Geotrichum		1.5
Absidia		1.5
	Torulaspora rosei	1.0
	Rhodotorula minuta	1.0
	Rhodotorula minta	1.0
Acremonium		1.0
Ulocladium		1.0
Trichophyton		1.0
Other filamentous fungi and yeast (0.5% each)	Subtotal	5.0
Unidentified organisms	Subtotal	1.5

[a] Percentages based on mean values.

Source: Adapted from Geldreich.[116]

the literature, a survey conducted by Kelley, Hall, and Paterson in 1997[202] concluded that only 3 studies examined filamentous fungi and yeasts in potable water in the U.K. and 24 in total worldwide. Recently, it has been suggested more studies need to be done to determine the frequency of fungi and the mycotoxins they produce in potable water.[203] A large diversity of fungi and yeast can be detected in potable water. This is shown in Table 8.6.

The numbers of fungi isolated in potable water have been documented at levels of 10^2 cfu per 100 ml.[24] Included in this amount are a number of different genera which have been identified from potable water in England and include *Cephalosporium* sp., *Verticillium* sp., *Trichoderma sporulosum, Nectria verdesceus, Phoma* sp., and *Phialophora sp.*[204,205] Generally, fungi within potable water do not constitute a public health concern but cause taste and odour problems at reported densities of 10 to 100 organisms per 100 ml.[204] However, one important fungi of human significance is *Aspergillus fumigatus* which has been identified from some 15 water supplies in Finland.[206] In general, the public health significance of these in potable water is unknown.

Yeasts do not have any significance as opportunistic pathogens. They have been detected at levels of 10^3 per ml in potable water[207] and have also been identified attached to pipe surfaces (Figures 8.1 to 8.3). An emerging opportunistic pathogen, being found in potable water, is *Candida albicans* which affects immunocompromised individuals.[208] Yeast cells have been shown to survive chlorine residuals as high as 1.2 mg per litre in potable water. This is owing to the thick and rigid cell wall presenting a barrier to chlorine.[110]

If we consider fungi and yeast in potable water, the problems they cause include[18]

1. Exerting a chlorine demand and also protecting bacterial pathogens from inactivation by chlorine.
2. Degrading joining compounds in distribution systems.
3. Forming humic substances which act as precursors of trihalomethanes.
4. Causing taste and odour problems.
5. Causing allergies and toxic reactions.

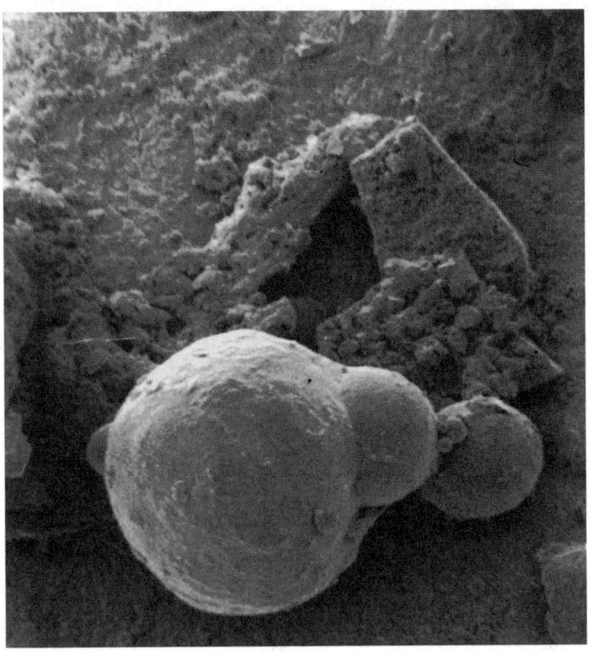

FIGURE 8.1 Evidence of a budding yeast attached to stainless steel in potable water.

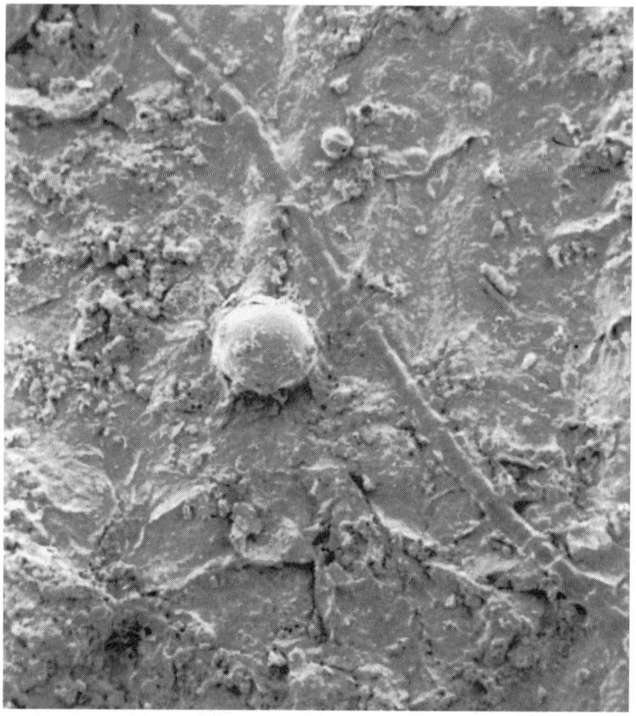

FIGURE 8.2 Evidence of yeast cells attached within a biofilm. Reprinted from *Water Research,* 32, Percival, S., Biofilms, mains water and stainless steel, 2187–2201, Copyright 1998, with permission from Elsevier Science.

8.7.11.12.2 Occurrence in biofilms

Fungi and yeasts have been found to colonise fire hydrants, elevated storage tanks, and attachment devices on buildings plumbing systems, for example, humidifiers and point of use devices.[209] They are also well known to be present in biofilms.[48,70,210] A very important public health concern as a result of growth within a biofilm is shown by the fact that colonisation onto pipe surfaces may exert a chlorine demand that would reduce available disinfectant residuals in the potable supply pipes. This would reduce the amounts necessary for the destruction of indicator organisms.

Incidents of biofilms incorporating yeasts are low. However, they have shown to be evident in old pipes, dead ends, and stagnant dead legs in potable water systems.[211]

8.7.11.13 Protozoa

A large number of protozoa have been identified within potable water. These include *Cryptosporidium*, *Giardia*, *Naegleria*, *Hartmanella*, and *Acanthamoeba*. Many protozoa (*Bodo*, *Voricella*, *Euplotes*) have also been identified in biofilms in water reservoirs.[18] These protozoa have been shown to withstand free chlorine residuals of 4 mg per litre or more for between 1 to 4 hours.[212] It is also well established that predatory protozoa incorporating opportunistic and pathogen bacteria allow for the protection of these microorganisms to high levels of free chlorine.[212]

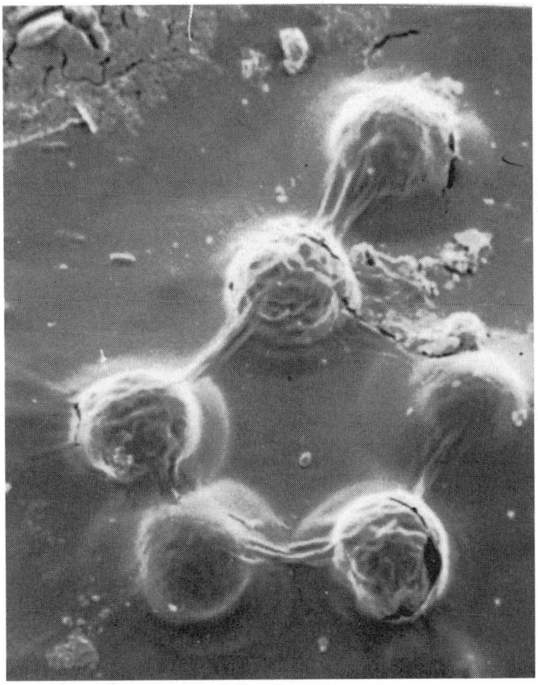

FIGURE 8.3 Evidence of yeast cells, attached by extracellular strands, in a potable water biofilm. Reprinted from *Water Research,* 32, Percival, S., Biofilms, mains water and stainless steel, 2187–2201, Copyright 1998, with permission from Elsevier Science.

8.7.11.14 *Cryptosporidium* spp.

8.7.11.14.1 General characteristics

Cryptosporidium was established as a significant infector of the human in the 1980s. It has been associated with boreholes[213] and are well known to infect pigs, mice, and lambs. Whilst there are about 20 species known to be evident, *Cryptosporidium parvum* is responsible for major infections in both animals and humans.[214]

Cryptosporidium consists of an oocyst normally 5 to 7 µm which provides it with very good protection in adverse environments. For the unsuspecting human, for infection to take place ingestion of the oocyst is necessary. Once inside the gut, the oocyst undergoes excitation resulting in the release of infective sporozoites which are capable of parasitizing epithelial cells within the gastrointestinal tract where they are taken into superficial parasitophorous vacuoles. It is here that the sporozoite matures and develops into a trophozoite which, in turn, divides to form a merozoite. These are then available to infect other epithelial cells leading to increases in the level of infection. In order to disseminate the merozoites form microgametes which develop into zygotes. The zygotes then mature into cysts. These cysts form the infective stage of the lifecycle and pass out in the faeces of infected individuals. The lifecycle of *Cryptosporidium parvum* is shown in Figure 8.4.

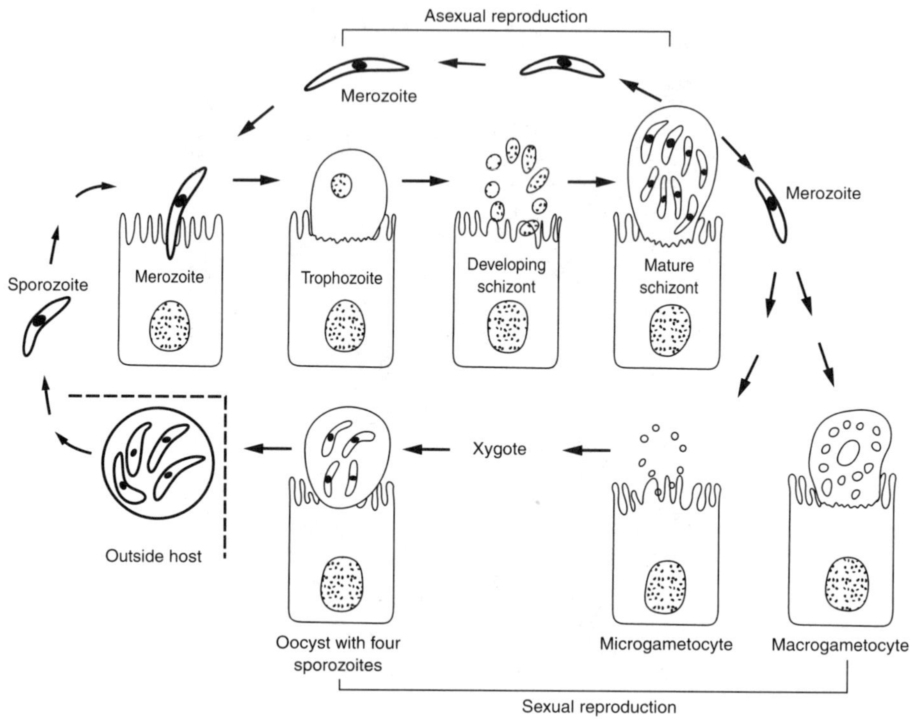

FIGURE 8.4 The lifecycle of Cryptosporidium. From *Waterborne Disease: Epidemiology and Ecology,* Hunter, P.R., Copyright © 1997. Reprinted by permission of Wiley-Liss, Inc., a subsidiary of John Wiley & Sons, Inc.

One hundred oocysts are regarded as the minimum infective dose known to cause human infection.[215,216] However, other animals have been shown to be infected by 1 to 30 oocysts.[217,218] Little information is known of the exact infectious dose size of oocysts to cause infection. It is thought to be lower than 100 oocysts.[219] Research suggests 10 oocytes will generate infection, however evidence shows that infection can occur with even a single oocyst.[215]

The symptoms associated with cryptosporidiosis include diarrhoea, which is generally watery and offensive, nausea, vomiting, weight loss, and fever. The incubation period following contamination with oocysts is 1 to 14 days and will continue indefinitely in immunocompromised patients. Symptoms usually last from 2 to 26 days.[162] The people most susceptible to the effects of *Cryptosporidium* include the elderly, young, and immunocomprised. In such individuals, nongastrointestinal illness, such as cholecystitis, hepatitis, and respiratory disease, may occur.[162]

8.7.11.14.2 Treatment of infection

There is no specific therapy available for the treatment of cryptosporidiosis. Management of the illness is supportive when it is necessary.

8.7.11.14.3 *Presence and outbreaks associated with potable water*

Many outbreaks associated with cryptosporidiosis have been recorded in the U.S. and Europe which raises serious concerns about the safety of potable water.[220-225] The exact public health significance of waterborne transmission of this organism, however, is unknown at present. This is particularly owing to the deficiencies in the methods for detecting, identifying, and counting oocysts, and it has hampered the understanding of the problems posed by *Cryptosporidium* in water supplies. Therefore, further work is necessary for the characterisation of these isolates and their virulence to man, particularly as it has been identified as the fourth major cause of diarrhoea in the U.K.[226] *Cryptosporidium* cysts also have been shown to be able to withstand normal chlorination procedures that are used in potable water systems.[227]

It has been well documented that *Cryptosporidium* is not efficiently removed or inactivated by traditional water treatment processes such as sand filtration or chlorination, although lime treatment has been shown to partially inactivate *Cryptosporidium* oocysts.[214] However, other research findings have suggested that normal chlorination procedures are generally effective.[92,228]

Juranek[229] has reported that *Cryptosporidium* infection can be transmitted from person to person through ingestion of contaminated water (drinking water and water used for recreational purposes) or food, from animal to person, or by contact with faecally contaminated environmental surfaces where this pathogen had formed part of a biofilm.

The number of cases of *Cryptosporidium* in England and Wales reported by the CDSC each year since 1990 has varied from about 4500 to 5200. Owing to the growing incidence of *Cryptosporidium,* particularly due to outbreaks in Oxford and Swindon in 1989, a group of experts was set up to give advice on the significance of *Cryptosporidium* in water supplies.[230,231] This group established that "normal treatment cannot prevent gross contamination of *Cryptosporidium* from entering water supplies." This is evident when we consider the total of 5705 cases of *Cryptosporidium* reported to the Public Health Laboratory Services and CDSC in 1995, compared to only 4432 cases recorded in 1994.[232]

Owing to the concerns being expressed with respect to *Cryptosporidium* in potable water systems, water companies are now setting up surveillance systems to monitor for it.

8.7.11.14.4 *Occurrence in biofilms*

Rogers and Keevil[233] have reported the retention and release of *Cryptosporidium* within a water distribution biofilm model for up to 6 weeks after inoculation into the biofilm. Such slow release may raise questions as to the surveillance period after an outbreak. Despite this work, the entrapment of *Cryptosporidium* spp. in biofilms is not well documented. However, studies have been carried out that have demonstrated that a significant number of *C. parvum* are able to persist and survive as oocysts within a biofilm for several weeks.[234] Therefore, oocysts which enter the potable water supply system may have the ability to reside in potable water biofilms. Possible sloughing of the biofilm could lead to infection even after water has been classified as safe. This and other types of waterborne disases may lead to occasional sporadic cases of infection which occur and whose source cannot be identified.

8.7.11.15 *Giardia* spp.

8.7.11.15.1 General characteristics

Giardia lamblia is now recognised as a significant cause of gastroenteritis in the U.S. inducing either mild or severe debilitating diseases.[235] They exist in two forms—the trophozoites (12 to 18 μm), which possess a concave sucking disc on the anterior end, and two falgella at the posterior end. The trophozoites are found adhered to the mucous membrane of the upper intestinal tract. This is responsible for causing disease. As the trophozoite travels down the gastrointestinal tract, it may develop into an oval cyst (9 to 12 μm long). These pass out in the faeces and form an infective agent. If they are ingested by a new host, they pass into the duodenum where they produce two daughter trophozoites which again colonise. It is known to have an incubation period of 5 to 25 days and produces abdominal pain, bloating, flatulence, and loose, pale, greasy stools. Symptoms usually last for 1 to 2 weeks, but they can be extended to months or even years. Consumption of 10 to 25 cysts generally causes an infection.

8.7.11.15.2 Treatment of infection

Fluid replacement is the most important mode of treatment for infection with this organism. However, specific antimicrobial agents can be given if infection is severe. The most effective antibiotics against this protozoa include metronidazole and tinidazole. However, in cases where these antibiotics have failed, mepacrine can be given.

8.7.11.15.3 Presence and outbreaks associated with potable water

Giardia has been estimated to be present in 81% of raw water and 17% of potable water supplies in the U.S.[236,237] With respect to potable water in the U.K., *Giardia* cysts have been found in approximately 48% of raw water and 23% of potable water supplies.[92] This may suggest why potable water outbreaks recorded in the U.K. are steadily increasing.[238,239]

Giardia cysts are very resistant to environmental conditions and once excreted, are capable of survival for long periods of time. It has also been found to be particularly resistant to normal chlorination processes used in water treatment.[240] The transmission of giardiasis is well documented[241] and is becoming a very important cause of waterborne disease.

8.7.11.15.4 Occurrence in biofilms

No literature is available as to the possible incorporation of *Giardia* sp. in biofilms. However, as with *Cryptosporidium*, entrapment within a biofilm seems feasible.

8.8 VIRUSES

8.8.1 General Characteristics

Viruses found within potable water systems which pose a public health threat are the so-called enteric viruses which are known to cause gastrointestinal problems. Of the 120 types of viruses known, the potential disease-causing viruses include calcivirus, rotavirus, astrovirus, Hepatitus A virus, Norwalk virus, and the small round viruses.[242]

Astroviruses have an incubation period of 1 to 4 days and cause acute gastroenteritis which lasts 2 to 3 days but does vary from 1 to 14 days. Calciviruses have a similar incubation period of 1 to 3 days and are also associated with acute gastroenteritis with symptoms lasting 1 to 3 days. Enteroviruses, which include poliovirus, coxsackieviruses, and echoviruses, produce a wide range of clinical symptoms following infection. These include febrile illness, respiratory problems, meningitis, herpangina, pleurodynia, conjunctivitis, myocardiopathy, diarrhoea, encephalitis, and ataxia. Once symptoms become apparent, these may last 1 to 3 days, but these do vary substantially. The rotaviruses are the main cause of gastroenteritis, transmitted by the faecal–oral route, particularly in children. These are divided into two groups, namely A and B. Group A has an incubation period of 1 to 3 days and is associated with acute gastroenteritis with predominant nausea and vomiting which may last for 12 to 48 hours. Group B has a similar incubation period and is also associated with acute gastroenteritis which lasts for 3 to 7 days. The Norwalk virus is also associated with acute gastroenteritis and vomiting which lasts for 12 to 48 hours.

Of major concern of all viruses are the ones which cause hepatitis. There are five major groups including Hepatitis A (transmitted mainly by water and faecal contaminated food), Hepatitis B (spread by personal contact or inoculation), and Hepatitis C, D, E, F, and G. Hepatitis A and E are the most predominate within potable water, particularly within the U.S.[243] Hepatitis A has an incubation period of 15 to 50 days and is associated with fever, malaise, jaundice, abdominal pain, and nausea which lasts for 1 to 2 weeks, with some cases lasting several months. The level of Hepatitis A virus excretion is short but high. It is known to be very resistant to chemical and physical agents.[244] Hepatitis E, however, has a slightly longer incubation period of 15 to 65 days and is associated with the same symptoms as group A but with a shorter duration of infection of 1 to 2 weeks. It is excreted briefly and at low concentrations. Epidemics of both viruses of this sort are generally in countries with poor sanitary conditions.

Viruses are excreted in much larger magnitudes than coliforms.[245] However, excreted viral numbers do not increase with their survival and infectivity rate dependant upon a moribund of factors including, in particular, the type of virus and abiotic factors it is exposed to, in particular, organic suspended matter.[246] It is generally found that suspended solids will provide a shield from biocides, extremes of temperature, and pH. However, most viruses are rapidly killed at temperatures above 50°C, suggesting a means of control.[247]

8.8.2 Treatment of Infection

With Hepatitis A and E, only supportive treatment is given. Itching can be controlled by antipruritic drugs and prolonged cholestatis steriods. For viral gastroenteritis, treatment takes the form of fluid replacement and supportive care. For enterovirus infection, no treatment is available. Treatment for adenoviral infection involves just supportive care with complete recovery evident in the majority of cases.

8.8.3 Presence and Outbreaks Associated with Potable Water

In recent years there has been concerns with respect to the isolation and detection of enteric viruses in chlorinated potable waters. Recent research in the U.S. suggests that there is low-level viral contamination of potable water resulting in sporadic viral infections amongst consumers.[248] However, in Europe there has been no evidence to substantiate this existence of a low-level transmission.[248]

Viruses of particular concern to the public's health include the Norwalk virus and its presence in potable water supplies amongst others.[249] This is possibly due to the fact that it is not affected by normal chlorination practices. Examples of outbreaks associated with the Norwalk virus have been documented.[18] Adenoviruses have also been associated with waterborne outbreaks. However, enteric adenoviruses are difficult to culture and so their prevalence within potable water is greatly underestimated. The large numbers of reported waterborne disease outbreaks which have no causative agent identified, possibly, are a result of enteric viruses, including adenoviruses.[250]

8.8.4 Occurrence in Biofilms

Despite the importance of biofilm development within potable water and their public health significance, very few studies have been carried out which are involved in the detection of enteric viruses within biofilms. Considering the levels of waterborne diseases evident in the U.K. and the world, it would be feasible to say viruses could become established within biofilms, ultimately providing protection from biocides. This would, therefore, suggest that biofilms within potable water systems have the potential to harbour enteric viruses. This concept has been hypothesised by vanden Bossche and Krietermeyer.[251] These researchers have detected various strains of type B coxsackievirus within biofilms colonising the outlet mains of a drinking water treatment plants. Quignon et al.[252] have also confirmed viral presence in potable water and biofilms.

8.9 INVERTEBRATES

8.9.1 Association with Potable Water

As well as containing bacteria, protozoa, fungi, and viruses, potable water may contain a number of invertebrates. These include nematodes, amphipods, copepods, and fly larva.[253] Whilst these have been identified in potable water, there is no

TABLE 8.7
Bacteria Isolated from Invertebrates Collected from the Drinking Water Distribution System of Worcester, MA, 1982–1984

Bacterium	Invertebrate type			
	Amphipod	Copepd	Fly larva	Nematode
Acinetobacter	+			
Achromobacter xylooxidans	+			
Aeromonas hydrophila			+	
Bacillus sp.	+			
Chromobacter violaceum	+			
Flavobacterium meningosepticum		+		
Moraxella sp.	+	+		
Pasteurella sp.		+		
Pseudomonas diminuta		+		
Pseudomonas cepacia			+	+
Pseudomonas fluorescens	+		+	
Pseudomonas maltophila	+			
Pseudomonas paucimobilis	+	+		
Pseudomonas vesicularis	+			
Serratia sp.	+			
Staphylococcus sp.	+			

+ = are present

Source: Adapted from Geldreich.[116]

evidence to suggest these organisms present any health risk at present. However, it is well documented that these organisms may harbor potential pathogens and also protect them from disinfection. Bacteria which have been isolated from invertebrates collected from potable water is shown in Table 8.7.

8.10 CONCLUSION

Water companies are legally obliged to supply potable water free of pathogenic agents or at a concentration which will not pose a potential risk to the health of the nation. Pathogens are generally absent from the public's drinking water. However, it is an emerging problem that the virulence of opportunistic bacterial pathogens (and also growing evidence of viral and protozoan pathogens) is becoming evident within potable waters. Despite waters being regarded as safe under the coliform concept, recent reports, together with self-medicated cases of waterborne diarrhoea, are showing reasons for concern. This is owing to the larger numbers of cases being attributed to both viral and protozoa infections and the validity of methods used to detect these organisms.

The reason for increases of diarrhoea owing to both viruses and protozoa over recent years are fivefold. These have been summarised by Gleeson and Gray[8] and include

1. Conventional sewage treatment has failed to remove them.
2. They are excreted in the faeces during illness.
3. They can survive in an environmentally robust form or they demonstrate resilience to inactivation in aquatic environments.
4. They are resistant to common disinfectants in potable water treatment.
5. They require low numbers to cause an infection in the host.
6. The problems in isolating and detecting pathogens.

Also, of serious concern within potable water is the virulence of opportunistic pathogens which, as well as being part of the normal planktonic heterotrophic bacterial population, are readily isolated as sessile populations removed from pipes transporting potable water systems. The problem associated with these organisms as mentioned by Gleeson and Gray[8] is not yet established, but there now seems to be an association between heterotrophic bacteria growing at 35°C and gastrointestinal illness.[254] Conclusions drawn recently within the U.K., however, have suggested that numbers of heterotrophic bacteria present in potable water are of little significance and constitute a nuisance to normal indicator organisms.[255] In this study, isolates were obtained from preflush and postflush potable water samples taken from both houses and multi-occupancy buildings from several geographically unrelated sites. Studies to determine the occurrence of toxigenic/virulence factors in the isolated heterotrophs were undertaken by assessing their potential to cause gastrointestinal upset. The study suggested that the frequency of isolation of heterotrophic species with the ability to cause gastrointestinal symptoms is too low to be of any public health concern. However, this area still remains a controversial topic of continuing debates.

In order for the establishment of zero levels of indicators and pathogens and low numbers of heterotrophic bacteria in potable water, changes are necessary in areas of water treatment and distribution. Also, more analytical methods need to be introduced on a larger scale. The use of alternative disinfectants such as chloramine, chlorine dioxide, and ozone will need critical evaluation in terms of their microbiological efficacy and their potential harmful by-products. Changes in the potable water distribution system, including engineering control measures, will be based on studies to determine the role of assimiable organic carbon and several other factors involved in the development of biofilms which may cause intermittent but persistent water quality problems. More efficient and cost-effective analytical methods will need to be developed for the detection of *Giardia* cysts, *Helicobacter pylori*, and *Legionella* in potable water environments.

The usage of periodic monitoring of water quality is very important, particularly with regards to determining the absence of pathogenic microorganisms. The detection of pathogens whose presence or absence in water is not related to faecal contamination or are more resistant to disinfection treatment necessitates invocation of such protocols. Also, the importance of newly emerged human pathogens transmitted through water should be recognised. Effective methods for their detection and removal need to be developed. Furthermore, with the prevalence of AIDS and changing demographics, human population groups may become vulnerable to those opportunistic pathogens that were traditionally regarded as environmental contaminants or merely as transient colonisers at unusual sites.

With the improved detection methodology, data about number and types of pathogens present in the aquatic environments will continue to grow and will greatly enhance our evaluation of the magnitude of microbial contamination of water sources. With better understanding, appropriate control measures can be instituted for prevention of health hazards associated with the presence of pathogens in water.

As pathogens and new emerging pathogens are evident and able to grow in potable water and biofilms, more research is needed to establish the health implications and risk associated with this mode of growth. This is particularly necessary as water considered safe under the present coliform concept is not a sure guarantee of the absence of pathogens in potable water and, therefore, health implications can be factorised by the presence of growth of these organisms within a biofilm.

Because of concern to the consumers, it is necessary that they apply certain recautionary procedures to protect themselves from public health issues related to water. As Geldreich[116] has mentioned, the most common problem evident in the home is the static water quality when water is first drawn in the morning. It is evident that microbial growth will occur in water during the overnight period. It is advised by Geldreich[116] to flush the tap water pipe for 30 seconds each morning before consuming the water. This is more of a concern during extended vacations.

8.11 REFERENCES

1. Ho, M., Glass, R. I., Pinsky, P. F., and Anderson, L., 1988, Rotavirus as a cause of diarrheal disease: a review of active survillance data, *Bull. World Health Organ.*, 60, 605.
2. Payment, P., Gramade, F., and Paquette, G., 1988, Microbiological and virological analysis of water from two water filtration plants and their distribution systems, *Can. J. Microbiol.*, 34, 1304.
3. Fawell, J. K. and Miller, D. G., 1992, Drinking water. A European comparison, *J. Inst. Water Environ. Manage.*, 6, 726.
4. Jones, K., 1994, Inside science: 73. Waterborne diseases, *New Sci.*, 143, 1.
5. World Health Organization, 1993, *Guidelines for Drinking Water Quality*, Vol. 1, *Recommendations*, 2nd ed., World Health Organization.
6. Environmental Protection Agency, 1992, National primary drinking water regulations — analytical techniques: coliform bacteria, final rule, *Fed. Regist.*, 57, 24744.
7. European Community, 1980, Council directive relating to the quality of water intended for human consumption (80/778/EEC), *Off. J. Eur. Commun.*, L229, 11.
8. Gleeson, C. and Gray, N., 1997, *The Coliform Index and Waterborne Disease: Problems of Microbial Drinking Water Assessment*, E and FN Spon, London.
9. Allen, M. J., Taylor, R. H., and Geldreich, E. E., 1980, The occurrence of microorganisms in water main encrustations, *J. Am. Water Works Assoc.*, 72, 614.
10. Block, J. C., Haudidier, K., Paquin, J. L., Miazga, J., and Levi, Y., 1993, Biofilm accumulation in drinking water distribution systems, *Biofouling*, 6, 333.
11. Walker, J. T., Mackerness, C. W., Roger, J., and Keevil, C. W., 1995, Biofilm — a haven for waterborne pathogens, in *Microbial Biofilms*, Lappin-Scott, H. M. and Costerton, J. W., Eds., Cambridge University Press, Cambridge.

12. Craun, G. F., 1988, Surface water supplies and health, *J. Am. Water Works Assoc.,* 80, 40.
13. LeChevallier, M. W., Lowry, C. D., and Lee, R. G., 1990, Disinfection of biofilms in model distribution system, *J. Am. Water Works Assoc.,* 82, 87.
14. Browning, J. R. and Ives, D. G., 1987, Environmental health and the water distribution system: a case history of an outbreak of giardiasis, *J. Inst. Water Environ. Manage.,* 1, 55.
15. Galbraith, N. S., Barnett, N. J., and Stanwell-Smith, R., 1987, Water and disease after Croydon: a review of waterborne associated disease in the United Kingdom (1937–1986), *J. Inst. Water Environ. Manage.,* 1, 7.
16. Stanwell-Smith, R., 1991, Recent trends in the epidemiology of waterborne disease, in *Proceedings of U.K. Symposium on Health Related Water Microbes,* Mavis, R., Alexander, L. M., Wyn-Jones, P., and Sellwood, J., Eds., 44.
17. Reasoner, D. J., 1992, Pathogens in drinking water — are there any new ones?, EPA, Washington, D.C.
18. Bitton, G., 1994, *Wastewater Microbiology,* Wiley, New York.
19. MacKerness, C. W., Colbourne, J. C., Dennis, P. J., Rachwal, A. J., and Keevil, C. W., 1992, Formation and control of coliform biofilms in drinking water distribution systems, in *Microbial Biofilms Formation and Control,* Denyer, S. P., Gorman, S.P., and Sussman, M., Eds., Soc. Appl. Bacteriol. Tech. Ser., 30, 217.
20. Lee, S. H., O'Conner, J. T., and Banerji, S. K., 1980, Biologically mediated corrosion and its effect on water quality in water systems, *J. Am. Water Works Assoc.,* 85, 111.
21. LeChevallier, M. W., Lowry, C. D., Lee, R. G., and Gibbon, L., 1993, Examining the relationship between iron corrosion and the disinfection of biofilm bacteria, *J. Am. Water Works Assoc.,* 82, 87.
22. Colin, F., Grapin, G., Cheron, J., Levi, Y., Pozzoli, E., Miazza, J., and Pascal, O., 1987, Etude de l'evolution del a qualite d'eau potable les reseaux de distribution, une approche et des myens mouveaux, *Tech. Sci. Munic. Eau,* 12, 556.
23. Camper, A. K., Hayes, J. T., Jones, W. L., and Zelver, N., 1993, Persistence of coliforms in mixed population biofilms, Proc. Water Qual. Tech. Conf., Miami American Water Works Association, Denver.
24. Holt, D., 1995, Challenge of controlling biofilms in water distribution systems, in *The Life and Death of Biofilm,* Wimpenny, J., Handley, P., Gilbert, P., Lappin-Scott, H., and Jones, M., Eds., Bioline, Cardiff, 161.
25. O'Conner, J. T. and Banerji, S. K., 1984, Biologically mediated corrosion and water quality deterioration in distribution systems, EPA Rep. 600/2/84/056, Cincinnati, OH, 442.
26. Olson, B. H. and Nagy, L. A., 1984, Microbiology of potable water, *Adv. Appl. Microbiol.,* 30, 73.
27. Percival, S. L., Knapp, J. S., Edyvean, R., and Wales, D. S., 1997, Biofilm development on 304 and 316 stainless steels in a potable water system, *J. Inst. Water Environ. Manage.,* 11, 289.
28. Shannon, A. M. and Wallace, W. M., 1944, The bacteria in a distribution system, *J. Am. Water Works Assoc.,* 36, 1356.
29. van Der Kooij, D., 1977, The occurrence of *Pseudomonas* spp. in surface water and tap water as determined on citrate media, *Antonie van Leeuwenhoek,* 43, 187.
30. Fransolet, G., Villers, G., and Masschelein, W. J., 1985, Influence of temperature on bacterial development in waters, *J. Int. Ozone Assoc.,* 7, 205.
31. Maki, J. S., La Croix, S. J., Hopkins, B. S., and Staley, J. T., 1986, Recovery and diversity of heterotrophic bacteria from chlorinated drinking waters, *Appl. Environ. Microbiol.,* 51, 1047.

32. LeChevallier, M. W., Babcock, T. M., and Lee, R. G., 1987, Examination and characterisation of distribution system biofilms, *Appl. Environ. Microbiol.,* 53, 2714.
33. Scarpino, P. V., Kellner, G. R., and Cook, H. C., 1987, The bacterial quality of bottled water, *J. Environ. Sci. Health,* 22, 357.
34. Baskerville, A., Fitzgeorge, R. B., Broster, M., Hambleton, P., and Dennis, P. J., 1981, Experimental transmission of legionnaires' disease by exposure to aerosols of *Legionella pneumophila, Lancet,* ii, 1389.
35. Manz, W., Szewzyk, U., Ericsson, P., Amann, R., Schleifer, K. H., and Stenstrom, T. A., 1993, *In situ* identification of bacteria in drinking water and adjoining biofilms by hybridization with 16S and 23S rRNA-directed fluorescent oligonucleotide probes, *Appl. Environ. Microbiol.,* 59, 2293.
36. Jess, J., Hight, N. J., Eptoin, H. A. S., Sigee, D. C., O'Neill, J. G., Meier, H., and Handley, P. S., 1997, Comparison of the API 20NE biochemical identification systems and partial 16s rRNA gene sequencing for the characterisation of culturable heterotrophs from a water distribution system biofilm, in *Biofilms: Community Interactions and Control,* Wimpenny, J., Handley, P., Gilbert, P., Lappin-Scott, H., and Jones, M., Eds., Bioline, Cardiff, 149.
37. Payment, P., Richardson, L., and Siemiatychi, J., 1991, Gastrointestinal health effects associated with the consumption of drinking water produced by point of use domestic reverse osmosis filtration units, *Appl. Environ. Microbiol.,* 57, 945.
38. Payment, P., 1997, Endemic gastrointestinal and respiratory disease: incidence, fraction attributable to tap water and costs to society, *Water Sci. Tech.,* 35, 7.
39. Agoustinos, M. T., Venter, S. N., and Kfir, R., 1992, Assessment of water quality problems due to microbial growth in drinking water distribution systems, Paper presented at the Int. Water Pollut. Res. Conf. Int. Symp., May 26–29, 1992, Washington, D.C.
40. LeChevallier, M. W., 1990, Coliform regrowth in drinking water: a review, *J. Am. Water Works Assoc.,* 82, 74.
41. Hutchinson, D., Weaver, R. H., and Scherago, M., 1943, The incidence and significance of microorganisms antagonistic to *E. coli, J. Bacteriol.,* 45, 29 Abs. G34.
42. Means, E. G. and Olson, B. H., 1981, Coliform inhibition by bacteriocin-like substances in drinking water distribution systems, *Appl. Environ. Microbiol.*, 42, 506.
43. LeChevallier, M. W., Seidler, R. J., and Evans, T. M., 1980, Enumeration and characterisation of SPC bacteria in chlorinated and raw water supplies, *Appl. Environ. Microbiol.,* 40, 922.
44. Oliveri, V. P., 1982, Bacterial indicators of pollution, in *Bacterial Indicators of Pollution,* Pipes, W.O., Ed., CRC Press, Boca Raton, FL, 21.
45. Geldreich, E. E., 1990, Microbiological quality control in distribution systems, in *Water Quality and Treatment*, Pontius, F.W., Ed., McGraw-Hill, New York.
46. Payment, P., Franco, E., and Siemiatycki, J., 1993, Absence of a relationship between health effects associated with the consumption and drinking water parameters, *Water Sci. Technol.,* 27, 137.
47. Geldreich, E. E., 1988, Coliform noncompliance nightmares in water supply distribution systems, in *Water Quality: A Realistic Perspective*, Univ. Michigan, College of Engineering; Michigan Section, American Water Works Association; Michigan Water Pollution Control Association; Michigan Department of Public Health, Lansing, MI.
48. Jarvis, W. R., 1990, Opportunistic pathogenic microorganisms in biofilms, Centers for Disease Control, Washington, D.C.
49. Clark, J. A., Burger, C. A., and Sabatinos, L. E., 1982, Characterisation of indicator bacteria in municipal raw water, drinking water and new main water samples, *Can. J. Microbiol.*, 28, 1002.

50. Ward, R. N., Wolfe, R. L., Justice, C. A., and Olson, B. H., 1986, The identification of gram-negative, non-fermentative bacteria from water: problems and alternative approaches to identification, *Adv. Appl. Microbiol.*, 31, 293.
51. Lindberg, R. B., 1974, Culture and identification of commonly encountered gram negative bacilli: *Pseudomonas, Klebsiella-Enterobacter, Serratia, Proteus* and *Providencia*, in *Opportunistic Pathogens*, Prier, J. E. and Friedman, H., Eds., University Park Press, Baltimore, MD.
52. Holder, I. A., 1977, Epidemiology of *Pseudomonas aeruginosa* in a burns hospital, in *Pseudomonas aeruginosa: Ecological Aspects and Patient Colonisation*, Young, V. M., Ed., Raven Press, New York.
53. Tinne, J. E., Gordon, A. M., Bain, W. H., and Mackey, W. A., 1967, Cross-infection by *Pseudomonas aeruginosa* as a hazard of intensive surgery, *Br. Med. J.*, 4, 313.
54. Schimpff, S. C., Greene, W. H., Young, V. M., and Wiernik, P. H., 1973, *Pseudomonas septicemia*: incidence, epidemiology, prevention and therapy in patients with advanced cancer, *Eur. J. Cancer*, 8, 449.
55. Hardalo, C. and Edberg, S. C., 1997, *Pseudomonas aeruginosa*: assessment of risk from drinking water, *Crit. Rev. Microbiol.*, 23, 47.
56. Gambassini, L., Sacco, C., Lanciotti, E., Burrini, D., and Griffini, O., 1990, Microbial quality of the water in the distribution system of Florence, *Aqua*, 39, 258.
57. Rogers, J., Dowsett, A. B., Dennis, P. J., Lee, J. V., and Keevil, C. W., 1994, Influence of plumbing materials on biofilm formation and growth of *Legionella pneumophila* in potable water systems, *Appl. Environ. Microbiol.*, 60, 1179.
58. Weber, G., Werner, H. P., and Matschnigg, H., 1971, *Pseudomas aeruginosa* in Trinkwasser als Todsursache bei Neugeborenen. Zentralbl. Bakteriol. Parasitenk. Infektionskr, *Hyg. I Abt. Orig.*, 216, 210.
59. Grundman, H., Kropec, A., Hartung, D., Berner, R., and Daschner, F., 1993, *Pseudomonas aeruginosa* in a neonatal intensive care unit: reservoirs and ecology of the nosocomial pathogens, *J. Infect. Dis.*, 168, 943.
60. Papapetropoulou, M., Rodopolou, G., and Giannoulaki, E., 1995, Improved glutamate-starch-penicillin agar for the isolation and enumeration of *Aeromonas hydrophila* from seawater by membrane filtration, *Pathol. Biol.*, 43, 622.
61. Olson, B. H., 1982, Assessment and implications of bacterial regrowth in water distribution systems, Project report EPA R-805680010, Order No. PB 82-249368, National Technical Information Service, Spingfield, VA.
62. Ridway, H. F. and Olson, B. H., 1982, Chlorine resistance patterns of bacteria from two drinking water distribution systems, *Appl. Environ. Microbiol.*, 44, 972.
63. Wolfe, R. L., Ward, N. R., and Olson, B. H., 1985, Inactivation of heterotrophic bacterial populations in finished drinking water by chlorine and chloramines, *Water Res.*, 19, 1393.
64. Franzblau, S. G., Jimenez, D. R., and Sinclair, N. A., 1985, A selective medium for the isolation of opportunistic flavobacteria from potable water, *J. Environ. Sci. Health*, 5, 583.
65. Favero, M. S., Peterson, N. J., Boyer, K. M., Carson, L. A., and Bond, W. W., 1974, Microbial contamination of renal dialysis systems and associated health risks, *Trans. Am. Soc. Artif. Intern. Organs*, 20A, 175.
66. Talley, M. W. and Alexander, M. R., 1989, Preventing potential health threats of stagnation in potable internal distribution systems, AWWA Conference Proceedings Part 2, 967.
67. Edmondson, E. B., Reinary, J. A., Pierce, A. K., and Sanford, J. P., 1966, Nebulisation equipment. A potential source of infection in gram-negative *Pneumonias, Am. J. Dis. Child.*, 111, 357.

68. Percival, S. L., Knapp, J. S., Edyvean, R., and Wales, D. S., 1998, Biofilm development on 304 and 316 stainless steels in a potable water system, *J. Inst. Water Environ. Manage.*, 11(4), 289.
69. Percival, S. L., Knapp, J. S., Wales, D. S., and Edyvean, R., 1998, The effects of the physical nature of stainless steel grades 304 and 316 on bacterial fouling, *Br. Corros. J.*, 33, 130.
70. Percival, S. L., Knapp, J. S., Wales, D. S., and Edyvean, R. G. J., 1999, The effect of turbulent flow and surface roughness on biofilm formation in drinking water, *J. Ind. Microbiol. Biotechnol.*, 22, 152.
71. Heron, D. S., McGonigle, B., Payer, M. A., and Baker, K. H., 1987, Attachment as a factor in the protection of *Enterobacter cloacae* from chlorination, *Appl. Environ. Microbiol.* 53, 178.
72. Armstrong, J. L., Calomiris, J. J., and Seidker, R. J., 1982, Selection of antibiotic resistant standard plate count bacteria during water treatment. *Appl. Environ. Microbiol.,* 44, 308.
73. Rice, E. W., Mesier, C. H., Johnson, C. H., and Reasoner, D. J., 1995, Occurrence of high-level aminoglycoside resistance in environmental isolates of enterococci, *Appl. Environ. Microbiol.,* 61, 374.
74. El-Zanfaly, H. T., Kassein, E. A., and Badr-Eldin, S. M., 1987, Incidence of antibiotic resistant bacteria in drinking water in Cairo, *Water Air Soil Pollut.*, 32, 123.
75. Geldreich, E. E., 1986, Control of microorganisms of public concern in water, *J. Environ. Sci.,* 29, 34.
76. Pipes, W. O., 1982, Introduction, in *Bacterial Indicators of Pollution,* Pipes, W. O., Ed., CRC Press, Boca Raton, FL, 1.
77. Waite, W. M., 1985, A critical appraisal of the coliform test, *J. Inst. Water Eng. Sci.,* 39, 341.
78. Rosenberg, M. L., Koplan, J. P., Wachsmuth, I. K., Wells, J. G., Gangarosa, E. J., Guerrant, R. L., and Sacks, P. A., 1977, Epidemic diarrhea at grater lake from enterotoxigenic escherichia coli, a large waterborne outbreak, *Ann. Intern. Med.,* 86, 714.
79. Harris, P., 1997, Legionella control in hot water, *Health Estate J.*, 51(6), 11.
80. Jones, I. G. and Roworth, M., 1996, An outbreak of *E. coli* 0157 and Campylobacteriosis associated with contamination of a drinking water supply, *Public Health,* 110, 227.
81. Dev, V. J., Main, M., and Gould, I., 1991, Waterborne outbreak of *Escherichia coli* 0157, *Lancet,* 337, 1412.
82. Geldreich, E. E., 1992, Visions of the future in drinking water water microbiology, *J. New Engl. Water Works Assoc.,* 56, 1.
83. Swerdlow, D. L., Woodruff, B. A., Brady, R. C., Griffin, P. M., Tippen, S., Donnell, Jr., H. D., Geldreich, E. E., Payne, B. J., Meyer, Jr., A., Wells, J. G., Greene, K. D., Bright, M., Bean, N. H., and Blake, P. A., 1992, A waterborne outbreak in Missouri of *Escherichia coli* 0157:H7 associated with bloody diarrhoea and death, *Ann. Intern. Med.,* 117, 812.
84. Drinking Water Inspectorate, 1998, Drinking water 1997—a report by the Chief Inspector Drinking Water Inspectorate, Dept. Environ., Transport, Regions, Stationary Office Publication.
85. Block, J. C., Mouteaus, L., Gatel, D., and Reasoner, D. J., 1997, Survival and growth of *E. coli* in drinking water distribution systems, in *Coliforms and E. coli: Problem or solution?,* Frickler, C. and Kay, D., Eds., Royal Society of Cambridge, Cambridge, 157.

86. Camper, A. K., Warnecke, K., Jones, W. L., and McFeters, G. A., 1998, Pathogens in model distribution system biofilms, AWWA Research Foundation Publication, AWWA, Denver, CO.
87. LeChevallier, M. W., Welsh, N. J., and Smith, D. B., 1996, Full scale studies of factors related to coliform regrowth in drinking water, *Appl. Environ. Microbiol.,* 62, 2201.
88. Fass, S., Dincher, M. L., Reasoner, D. J., Gatel, D., and Block, J. C., 1996, Fate of *Escherichia coli* experimentally injected into a drinking water distibution pilot system, *Water Res.*, 30, 2215.
89. Wiedenmann, A., Braun, M., and Botzenhart, K., 1997, Evaluation of the disinfection potential of lower chlorine concentrations in tap water using immobilised *Enterococcus faecium* in a continous flower device, *Water Sci. Technol.,* 35, 77.
90. Sibille, I., Sime-Ngando, T., Mathieu, L., and Block, J. C., 1998, Protozoan bacterivory and *Escherichia coli* survival in drinking water distribution systems, *Appl. Environ. Microbiol.,* 64, 197.
91. McMath, S. M., Sumpter, C., Holt, D. M., Delanoue, A., and Chamberlain, A. H. L., 1999, The fate of environmental coliforms in model water distribution system, *Lett. Appl. Microbiol.*, 28, 93.
92. Gray, N. F., 1994, *Drinking Water Quality: Problems and Solutions,* Wiley, Chichester.
93. LeChevallier, M. W., Cawthon, C. P., and Lee, R. G., 1988, Factors promoting survival of bacteria in chlorinated water supplies, *Appl. Environ. Microbiol.,* 54, 649.
94. LeChevallier, M. W., Cawthon, C. P., and Lee, R. G., 1988, Inactivation of biofilm bacteria, *Appl. Environ. Microbiol.,* 54, 2492.
95. Geldreich, E. E., Nash, H. D., Reasoner, D. J., and Taylor, R. H., 1972, The necessity of controlling bacterial populations in potable waters: community water supply, *J. Am. Water Works Assoc.,* 64, 596.
96. Camper, A. K., Jones, W. L., and Hayes, J. T., 1996, Effects of growth conditions and substratum composition on the persistence of coliforms in mixed-population biofilms, *Appl. Environ. Microbiol.,* 62, 4014.
97. Earnhardt, K. B., 1980, Chlorine resistant coliforms—the Muncie, Indiana, experience, Proc. AWWA Water Qual. Tech. Conf., Miami Beach, FL, American Water Works Association, Denver, CO.
98. Lowther, E. D. and Moser, R. H., 1984, Detecting and eliminating coliform regrowth, Proc. AWWA Water Qual. Tech. Conf., Denver, CO, American Water Works Association, Denver, CO.
99. Olivieri, V. P., Bakalian, A. E., Bossung, K. W., and Lowther, E. D., 1985, Recurrent coliforms in water distribution systems in the presnce of free residual chlorine, in *Water Chlorination, Chemistry, Environmental Impact and Health Effects,* Jolley R. L., Ed., Lewis Publishers, Chelsea, MI, 651.
100. Smith, H. V., Patterson, W. J., Hardie, R., Greene, L. A., Benton, C., Tulloch, W., Gilmour, R. A., Girdwood, R. W. A., Sharp, J. C. M., and Forbes, G. I., 1989, An outbreak of cryptosporidiosis caused by post-treatment contamination, *Epidemiol. Infect.,* 103, 703.
101. Wierenga, J. T., 1985, Recovery of coliforms in the presence of free chlorine residual, *J. Am. Water Works Assoc.,* 78, 83.
102. Hudson, L. D., Hankins, J. W., and Battaglia, B., 1983, Coliforms in a water distribution system: a remedial approach, *J. Am. Water Works Assoc.,* 75, 564.
103. Lightfoot, N., 1997, The health significance of heterotrophic bacteria in drinking water, in *The Microbiological Quality of Water*, Sutcliffe, D. W., Ed., Freshwater Biology Association. Titus Wilson and Son, Kendal, U.K., 35.

104. Aleksic, S. and Bockemuhl, J., 1988, Serological and biochemical characteristics of 416 *Yersinia* strains from well water and drinking water plants in the Federal Republic of Germany: lack of evidence that these strains are of public health importance, *Zentralbl. Bakteriol. Mikrobiol. Hyg.* (*B*), b185, 527.
105. Eden, K. V., Rosenberg, M. L., Stoopler, M., Wood, B. T., Highsmith, A. K., Skaliy, P., Wells, J. G., and Feeley, J. C., 1977, Waterborne gastrointestinal illness at a ski resort, *Pub. Health Rep.,* 92, 245.
106. Ostroff, S. M., Kapperud, G., Hutwagner, L. C., Nesbakken, T., Bean, N. H., Lassen, J., and Tauxe, R. V., 1994, Sources of sporadic *Yersinia enterocolitica* infections in Norway: a prospective case-control study, *Epidemiol. Infect.,* 112, 33.
107. Goslee, S. and Wolinsky, E., 1976, Water as a source of potentially pathogenic mycobacteria, *Am. Rev. Respir. Dis.,* 113, 287.
108. du Molin, G. C., Sherman, I. H., Hoaglin, D. C., and Stottmeier, K. D., 1985, *Mycobacterium avium* complex, and emerging pathogen in Massachusetts, *J. Clin. Microbiol.,* 22, 9.
109. Jenkins, P. A., 1991, *Mycobacterium* in the environment, *J. Appl. Bacteriol.,* Symposium Supplement, 137.
110. Engelbrecht, R. S., Severin, B. F., Masarek, M. T., Faroog, S., Lee, S. H., Haas, C. M., and Lalchandani, A., 1977, New microbial indicators of disinfection efficiency, EPA-600/2-77-052, Environmental Protection Agency, Cincinnati, OH.
111. Shular, J. A., 1978, The occurrence of indicator organisms in the Decatur, Illinois South Water treatment plant and in one branch of the distribution system, M.S. Special Report, Dep. Civil Engineering, Univ. Illinois, Urbana-Champaign.
112. Surucu, F. and Haas, C. N., 1976, Inactivation of new indicator organisms of disinfection efficiency. Part I. Free available chlorine species kinetics, presented at the 96th annual meeting, American Water Works Association, New Orleans.
113. Pelletier, P. A., duMoulin, G. C., and Stottmeir, K. D., 1988, Mycobacteria in public water supplies: comparative resistance to chlorine, *Microbiol. Sci.,* 5, 147.
114. Farooq, S., 1976, Kinetics of inactivation of yeasts and acid fast organisms with ozone, Ph.D. thesis, Univ. Illinois, Dept. Civil Engineering, Urbana-Champaign.
115. Severin, B. F., 1976, Inactivation of new indicator organisms of disinfection efficiency. Part II. Combined chlorine as chloramines, presented at the 96th annual meeting, American Water Works Association, New Orleans.
116. Geldreich, E. E., Ed., 1996, *Microbial Quality of Water Supply in Distribution Systems,* Lewis Publishers, New York.
117. du Molin, G. C. and Stottmeier, K. D., 1986, Waterborne mycobacteria: an increasing threat to health, *Am. Soc. Microbiol. News,* 1986, 525.
118. Panwalker, A. P. and Fuhse, E., 1986, Nosocomial *Mycobacterium gordonae* pseudo infections from contaminated ice machines, *Infect. Contr.,* 7, 67.
119. Schulze-Robbecke, R., Janning, B., and Fischeder, R., 1992, Occurrence of mycobacteria in biofilm samples, *Tuberc. Lung Dis.,* 73, 141.
120. Schulze-Robbecke, R. and Fischeder, R., 1992, Mycobacteria in biofilms, *Zentralbl. Hyg. Umweltmed.,* 188, 385.
121. Hall-Stoodley L., Keevil, C. W., and Lappin-Scott, H. M., 1999, *Mycobacterium fortuitum* and *Mycobacterium chelonae* biofilm formation under high and low nutrient conditions, *J. Appl. Microbiol. Symp. Suppl.,* 85, 60S.
122. Payment, P., Eduardo, F., Richardson, L., Siemiatychi, J., Dewar, R., Edwards, M., and Franco, E., 1991, A randomised trial to evaluate the risk of gastrointestinal disease due to the consumption of drinking water meeting currently accepted microbiological standards, *Am. J. Pub. Health,* 81, 703.

123. Jones, K. and Bradshaw, S. B., 1996, Biofilm formation by the enterobacteriaceae: a comparison between *Salmonella enteritidis*, *Escherichia coli* and a nitrogen-fixing strain of *Klebsiella pneumoniae, J. Appl. Bacteriol.*, 80, 458.
124. Hood, S. K. and Zottola, E. A., 1997, Adherence to stainless steel by foodborne microorganisms during growth in model food systems, *Int. J. Food Microbiol.*, 37, 145.
125. Korber, D. R., Choi, A., Wolfaardt, G. M., Ingham, S. C., and Caldwell, D. E., 1997, Substratum topography influences susceptibility of *Salmonella enteritidis* biofilms to trisodium phosphate, *Appl. Environ. Microbiol.*, 63, 3352.
126. Blaser, M. J., Pollard, R. A., and Feldman, R. A., 1984, *Shigella* infections in the United States 1974–1980, *J. Infect. Dis.*, 147, 771.
127. Samonis, G., Elting, L., and Skoulika, E., 1994, An outbreak of diarrhoeal disease attributed to *Shigella sonnei, Epidemiol. Infect.*, 112, 235.
128. Merson, M. H., Tenny, J. H., Meyers, J. D., Wood, B. T., Wells, J. G., Rymzo, W., Cline, B., De Witt, W. E., Skaliy, P., and Mallison, F., 1975, Shigellosis at sea: an outbreak aboard a passenger cruise ship, *Am. J. Epidemiol.*, 101, 165.
129. Baine, W. B., Herron, C. A., Bridson, K., Baker, Jr., W. H., Lindell, S., Mallison, G. F., Wells, J. G., Martin, W. T., Kosuri, M. R., Carr, F., and Voelker, Sr., E., 1975, Waterborne outbreak at a public school, *Am. J. Epidemiol.*, 101, 323.
130. Weissman, J. B., Craun, G. F., Lawrence, D. N., Pollard, R. A., Saslaw, M. S., and Gangarosa, E. J., 1976, An epidemic of gastroenteritis traced to a contaminated public water supply, *Am. J. Epidemiol.*, 103, 391.
131. Skirrow, M. B., 1977, *Campylobacter enteritis*: a "new" disease, *Br. Med. J.*, ii, 9.
132. Bolton, F. J., Hinchliffe, P. M., Coates, D., and Robertson, L., 1982, A most probable number method for estimating small numbers of *Camplyobacter* in water, *J. Hyg.*, 89, 185.
133. Benson, A., 1995, *Control of Communicable Disease,* American Public Health Association.
134. Andersson, Y. and Stenstrom, T. A., 1986, Waterborne outbreaks in Sweden causes and etiology, *Water Sci. Technol.*, 18, 185.
135. Blaser, M. J., Taylor, D. N., and Feldman, R. A., 1982, Epidemiology of *Campylobacter jejuni* infections, *Epidemiol. Rev.*, 5, 157.
136. Blaser, M. J. and Cody, H. J., 1986, Methods for isolating *Campylobacter jejuni* from low-turbidity water, *Appl. Environ. Microbiol.*, 51, 312.
137. Skirrow, M. B. and Blaser, M. J., 1992, Clinical and epidemiological considerations, in *Campylobacter jejuni, Current Status and Future Trends,* Nachamkin, I., Blaser, M. J., and Tomkins, L. S., American Society for Microbiology, Washington, D.C., 3.
138. Carter, A. M., Pacha, R. E., Clark, G. W., and Williams, E. A., 1987, Seasonal occurrence of *Campylobacter* spp. in surface water and their correlation with standard indicator bacteria, *Appl. Environ. Microbiol.*, 53, 523.
139. Buswell, C. M., Herlihy, Y. M., Lawrence, L. M., McGuiggan, J. T. M., Marsh, P. D., Keevil, C. W., and Leach, S. A., 1998, Extended survival and persistence of *Campylobacter* spp. in water and aquatic biofilms and their detection by immunofluorescenet antibody and rRNA staining, *Appl. Environ. Microbiol.*, 64, 733.
140. Mead, G. C., Hudson, W. R., and Hinton, M. H., 1995, Effect of changes in processing to improve hygiene control on contamination of poultry carcasses with campylobacter, *Epidemiol. Infect.*, 115, 495.
141. Dennis, P. J., Green, D., and Jones, B. P. C., 1984, A note on the temperature tolerance of legionella, *J. Appl. Bacteriol.*, 56, 349.

142. Tiefenbrunner, F. A., Arnold, M., Dierichi, P., and Emde, K., 1993, Occurrence and distribution of *Legionella pneumophila* in water systems of Central European private homes, in *Legionella Current Status and Emerging Perspectives,* Barbaree, J. M., Brieman, R. F., and Dufour, A. P., Eds., American Society of Microbiology, Washington, D.C., 235.
143. Tobin, J. O. H., Barrtlett, C. L. R., Waitkins, S. A., Barrows, G. I., Macrae, A. D., Taylor, A. G., Fallon, R. J., and Lynch, F. R., 1981, Legionnaires' disease: further evidence to implicate water storage and distribution systems as sources, *Br. Med. J.,* 282, 573.
144. Best, M., Goetz, A., and Yu, V. L., 1984, Heat eradication measures for control of nosocomial Legionnaires' disease. Implementation, education, and cost analysis, *Am. J. Infect. Control,* 12(1), 26.
145. Dennis, P. J., Taylor, J. A., Fitzgeorge, R. B., Barlett, C. L. R., and Barrow, G. I., 1982, *Legionella pnemophila* in water plumbing systems, *Lancet* i, 949.
146. **Anon.,** 1990, The control of legionellosis including legionnaires' disease, Health and Safety Booklet HS(G) 70, Her Majesty's Stationary Office, London.
147. Comitte of Enquiry, 1987, Second report into the outbreak of Legionnaires' disease in Stafford in April, HMSO, London.
148. Barnstein, N., Vieilly, C., Nowicki, M., Paucod, J. C., and Fleurette, J., 1986, Epidemiological evidence of Legionellosis transmission through domestic hot water supply systems and possibilities of control, *Isr. J. Med. Sci.,* 22, 655.
149. Lee, J. V. and West, A. A., 1991, Survival and growth of Legionella species in the environment, *Soc. Appl. Bacteriol. Symp. Ser.,* 20, 121S.
150. Kilvington, S. and Price, J., 1990, Survival of *Legionella pneumophila* within cysts of *Acanthamoeba polyphaga* following chlorine exposure, *J. Appl. Bacteriol.,* 68, 519.
151. Skaliy, P., Thompson, T. A., Gorman, G. W., Morris, G. K., McEachern, H. V., and Mackel, D. C., 1980, Laboratory studies of disinfectants against *Legionella pneumophila, Appl. Environ. Microbiol.,* 40, 697.
152. Wadowsky, R. M., Yee, R. B. and Megmar, L., 1982, Hot water systems as a source of *Legionella pneumophila* in hospital and non-hospital plumbing fixtures, *Appl. Environ. Microbiol.,* 43, 1104.
153. Farrell, I. D., Barker, J. E., Miles, E. P., and Hutchison, J. G., 1990, A field study of the survival of *Legionella pneumophila* in a hospital hot-water system, *Epidemiol. Infect.,* 104, 381.
154. Barker, J., Brown, M. R., Collier, P. J., Farrell, I., and Gilbert, P., 1992, Relationship between *Legionella pneumophila* and *Acanthamoeba polyphaga*: physiological status and susceptibility to chemical inactivation, *Appl. Environ. Microbiol.,* 58, 2420.
155. Wasowsky, R. M. and Yee, R. B., 1985, Effect of non-Legionellaceae bacteria on the multiplication of *Legionella pneumophila* in potable water, *Appl. Environ. Microbiol.,* 49, 1206.
156. Rogers, J. and Keevil, C. W., 1992, Immunogold and fluorescein immuno-labelling of *Legionella pneumophila* within an aquatic biofilm visualised using episcopic differential interference contrast micrscopy, *Appl. Environ. Microbiol.,* 58, 2326.
157. Armon, R., Starosvetzky, T., Arbel, T., and Green, M., 1997, Survival of *Legionella pneumophila* and *Salmonella typhimurium* in biofilm systems, *Water Sci. Technol.,* 35, 293.
158. Gracey, M., Burke, V., and Robinson, J., 1982, Aeromonas associated gastroenteritis, *Lancet,* 2, 1304.
159. Altwegg, M. and Geiss, H. K., 1989, *Aeromonas* as a human pathogen, *CRC Crit. Rev. Microbiol.,* 16, 253.

160. Namdari, H. and Bottone, E. J., 1990, Microbiologic and clinical evidence supporting the role of *Aeromonas caviae* as a paediatric enetric pathogen, *J. Clin. Microbiol.*, 28, 837.
161. Merino, S., Rubries, X., Knochel, S., and Tomas, J. M., 1995, Emerging pathogens: *Aeromonas* spp., *Int. J. Food Microbiol.*, 28, 157.
162. Hunter, P. R., 1997, *Waterborne Disease: Epidemiology and Ecology*, John Wiley & Sons, London.
163. Janda, J. M. and Duffey, P. S., 1988, Mesophilic aeromonads in human disease: current taxonomy, laboratory identification and infectious disease spectrum, *Rev. Infect. Dis.*, 10, 980.
164. Morgan, D. R., Johnson, P. C., DuPont, H. L., Salterwhite, T. K., and Wood, L. V., 1985, Lack of correlation between known virulence properties of *Aeromonas hydrophila* and enteropathogenicity for humans, *Infect. Immunol.*, 50, 62.
165. Havelaar, A. H., Versteegh, J. F. M., and During, M., 1990, The presence of *Aeromonas* in drinking water supplies in the Netherlands, *Zentralbl. Hyg.*, 190, 361.
166. Burke, V., Robinson, J., Gracey, M., Peterson, D., and Partridge, K., 1984, Isolation of *Aeromonas hydrophila* from a metropolitan water supply: seasonal correlation with clinical isolates, *Appl. Environ. Microbiol.*, 48, 361.
167. Schbert, R. H., 1991, Aeromonads and their significance as potential pathogens in water, *J. Appl. Bacteriol.*, Symposium Supplement, 70, 131.
168. Hanninen, M. L. and Siitonen, A., 1995, Distribution of *Aeromonas* phenospecies and genospecies among strains isolated from water, foods or from human clinical samples, *Epidemiol. Infect.*, 115, 39.
169. Krovacek, K., Faris, A., Baloda, S. B., Lindberg, T., Peterz, M., and Mansson, I., 1992, Isolation and virulence profiles of *Aeromonas* spp. from different municiple drinking water supplies in Sweden, *Food Microbiol.*, 9, 215.
170. Fewtrell, L., Godfree, A., Jones, F., Kray, D., Salmon, R. L., and Wyer, M. D., 1992, Health effects of white-water canoeing, *Lancet*, 339, 1587.
171. Burke, V., Robinson, J., and Gracey, M., 1984, Isolation of *Aeromonas hydrophila* from a metropolitan water supply: seasonal correlation with clinical isolates, *Appl. Environ. Microbiol.*, 48, 361.
172. Araujo, R. M., Arribas, R. M., and Pares, R., 1991, Distribution of *Aeromonas* species in water with different levels of pollution, *J. Appl. Bacteriol.*, 71, 182.
173. Fiorentino, P. G., 1995, Isolamento di *Aeromonas* spp. da acque potabili in provincia di Firenze, *Inquinamento*, 27, 46.
174. Legani, P., Leoni, E., Soppelsa, F., and Burigo, R., 1998, The occurrence of Aeromonas species in drinking water supplies of an area of the Dolomite Mountains, Italy, *J. Appl. Microbiol.*, 85, 271.
175. Millership, S. E. and Chattopadhyay, B., 1985, *Aeromonas hydrophila* in chlorinated water supplies, *J. Hosp. Infect.*, 6, 75.
176. Lightfoot, N. F., Nichols, G. L., de Louvois, J., and Adak, R., 1995, Prevalence study of specific bacterial infections and the potential risk of transmission via drinking water, United Kingdom Research Limited, London.
177. Massa, S., Armuzzi, R., Tosques, M., Canganella, F., and Trovatelli, L. D., 1999, Note: susceptibility to chlorine of *Aeromonas hydrophila* strains, *J. Appl. Microbiol.*, 86, 169.
178. Ozbac, Z. Y. and Aytac, S. A., 1994, Effect of chlorine on growth and survival of *Aeromonas hydrophila* and enteropathogenicity for human, *Infect. Immun.*, 50, 62.

179. LeChevallier, M. W., Evans, T. M., Seidler, R. J., Daily, O. P., Merrell, B. R., Rollins, D. M., and Joseph, S. W., 1982, *Aeromonas sobria* in chlorinated drinking water supplies, *Microbial. Ecol.,* 8, 325.
180. Burke, V. J., Robinson, M., and Gracey, M., 1984, Isolation of *Aeromonas* spp. from an unchlorinated domestic water supply, *Appl. Environ. Microbiol.,* 48, 367.
181. Havelaar, A. H., Versteegh, J. F. M., and During, M., 1990, The presence of *Aeromonas* in drinking water supplies in the Netherlands, *Zentralbl. Hyg. Umweltmediz.,* 190, 236.
182. van Der Kooij, D., 1991, Nutritional requirements of aeromonads and their multiplication in drinking water, *Experientia*, 47, 444.
183. Knochel, S. and Jeppesen, C., 1990, Distribution and characteristics of Aeromonas in food and drinking water in Denmark, *Int. J. Food Microbiol.,* 10, 317.
184. Stelzer, W., Jacon, J., Feuerpfeil, I., and Schulze, E., 1992, The occurrence of aeromonads in a drinking water supply system, *Zentralbl. Mikrobiol.,* 147, 231.
185. Aulicino, F. A. and Orsini, P., 1994, La ricrescita batterica in reti idriche e aspetti igienico-sanitari, *Ann. Igiene,* 6, 861.
186. Holmes, P. and Niccolls, L. M., 1995, Aeromonads in drinking water supplies: their occurrence and significance, *J. Inst. Water Environ. Manage.,* 5, 464.
187. Bhattacharya, S. K., Bhattacharya, M. K., and Nair, G. B., 1993, Clinical profile of acute diarrhoeal cases infected with new epidemic strains of *Vibrio cholera* 0139: designation of the disease as cholera, *J. Infect.*, 27, 11.
188. Bodhidatta, L., Echeverria, P., and Hoge, C. W., 1995, *Vibrio cholera* 0139 in Thailand in 1994, *Epidemiol. Infect.*, 114, 71.
189. Marshall, B. J. and Warren, J. R., 1984, Unidentified curved bacilli in the stomach of patients with gastritis and peptic ulceration, *Lancet,* I, 1311.
190. Reilly, J. P., 1998, Overview of *H. pylori* in peptic ulcer disease, *Am. J. Managed Care,* 4, S247.
191. West, P. A., Millar, M. R., and Tompkins, D. S., 1992, Effects of physical environment on survival of *Helicobacter pylori, J. Clin. Pathol.,* 45, 228.
192. Hulten, K., Han, S. W., Enroth, H., Klein, P. D., Opekun, A. R., Evans, D. G., Engstrand, L., Graham, D. Y., and El-Zaatari, F. A., 1996, *Helicobacter pylori* in the drinking water in Peru, *Gastroenterology,* 110, 1031.
193. Mara, D. D. and Clapham, D., 1997, Water-related carcinomas: environmental classification, *J. Enviorn. Eng.—ASCE,* 123, 416.
194. Johnson, C. H., Rice, E. W., and Reasoner, D. J., 1997, Inactivation of *Helicobacter pylori* by chlorination, *Appl. Environ. Microbiol.,* 63, 4969.
195. Fan, X. G., Chua, A., Li, T. G., and Zeng, Q. S., 1998, Survival of *Helicobacter pylori* in milk and tap water, *J. Gastroenterol. Hepatol.*, 13, 196.
196. Mackay, W. G., Gribbon, L. T., Barer, M. R., and Reid, D. C., 1998, Biofilms in drinking water systems—a possible reservoir for *Helicobacter pylori, Water Sci. Technol.,* 38, 181.
197. Mackay, W. G., Gribbon, L. T., Barer, M. R., and Reid, D. C., 1999, Biofilms in drinking water systems—a possible reservoir for *Helicobacter pylori, J. Appl. Microbiol.,* Symposium Supplement, 85, 52S.
198. Nagy, L. A. and Olson, B. H., 1982, The occurrence of filamentous fungi in drinking water distibution systems, *Can. J. Microbiol.* 28, 667.
199. Niemi, R. M., Knuth, S., and Lundstrom, K., 1982 Actinomycetes and fungi in surface waters and potable water, *Appl. Environ. Microbiol.*, 43, 378.
200. Rosenzweig, W. D., Minnigh, H. A., and Pipes, W. O., 1983, Chlorine demand and inactivation of fungal propagules, *Appl. Environ. Microbiol.*, 45, 182.

201. Rosenzweig, W. D. and Pipes, W. O., 1989, Presence of fungi in drinking water, in *Biohazards of Drinking Water Treatment*, Larson, R. A., Ed., Lewis, Cheslsea, 85.
202. Kelley, J., Hall, G., and Paterson, R. R. M., 1997, The significance of fungi in water distribution systems, Report DW-01/F for U.K. Water Industry Research Limited, International Mycological Institute, Egham.
203. Kinsey, G. C., Paterson, R. R., and Kelley, J., 1999, Methods for the determination of filamentous fungi in treated and untreated waters, *J. Appl. Microbiol.,* Symposium Supplement, 85, 214S.
204. Burman, N. P., 1965, Taste and odour due to stagnation and local warming in long lengths of piping, *Soc. Water Treat. Exam.*, 14, 125.
205. Bays, L. R., Burman, N. P., and Lavis, W. M., 1970, Taste and odour in water supplies in Great Britain: a study of the present position and problems for the future, *J. Soc. Water Treat. Exam.*, 19, 136.
206. Gutman, A. A., 1985, Allergens and other factors important in atopic disease, in *Allergic Diseases: Diagnosis and Management*, 3rd ed., Patterson, R., Ed., Lippincott, New York.
207. O'Connor, J. T., Hash, L., and Edwards, A. B., 1975, Deterioration of water quality in distribution systems, *J. Am. Water Works Assoc.,* 67, 113.
208. Ahearn, D. G., 1974, Identification and ecology of yeasts of medical importance, in *Opportunistic Pathogens,* Prier, J. E. and Friedman, H., Eds., University Park Press, Baltimore, MD.
209. Gerlach, W., 1972, Fusarien aus Trinkwasserleitungen, *Ann. Agric. Fenn.*, 11, 298.
210. Nagy, L. A. and Olson, B. H., 1985, Occurrence and significance of bacteria, fungi and yeasts associated with distribution pipe surfaces, *Proc. Water Qual. Tech. Conf.,* Houston, TX.
211. Hinzelin, F. and Block, J. C., 1985, Yeast and filamentous fungi in drinking water, *Environ. Technol. Lett.*, 6, 101.
212. King, C. H., Shotte, E. B., Wooley, R. E., and Porter, K. G., 1988, Survival of coliforms and bacterial pathogens within protozoa during chlorination, *Appl. Environ. Microbiol.*, 54, 3023.
213. Morgan, D., Allaby, M., Crook, S., Casemore, D., Healing, T. D., Soltanpoor, N., Hill, S., and Hooper, W., 1996, Waterborne cryptosporidiosis associated with a borehole supply, *Commun. Dis. Rep.,* 5, r93.
214. Rose, J. B., 1990, Emmerging issues for the microbiology of drinking water, *Eng. Manage.*, 90, 23.
215. Blewett, D. A., Wright, J. J., Casemore, D. P., Booth, N. E., and Jones, C. E., 1993, Infective dose size studies on *Cryptosporidium parvum* using gnotobiotic lambs, *Water Sci. Technol.,* 27, 61.
216. Peeters, J. E., van Opdenbosch, E., and Glorieux, B., 1989, Effect of disinfection of drinking water with ozone or chlorine dioxide on survival of *Cryptosporidium parvum* oocysts, *Appl. Environ. Microbiol.,* 55, 1519.
217. Rose, J. B., 1988, Occurrence and significance of cryptosporidium in water, *J. Am. Water Works Assoc.,* 88, 53.
218. Guerrant, R. L., 1997, Cryptosporidiosis are emerging, highly infectious threat, *Emerg. Infect. Dis.,* 3(1), 51.
219. Casemore, D., 1990, Epidemiological aspects of human cryptosporidiosis, *Epidemiol. Infect.*, 104, 1.
220. Hayes, E. B., Matte, T. D., O'Brien, T. R., McKinley, T. W., Logsdon, G. S., Rose, J. B., Ungar, B. L. P., Word, D. M., Pinsky, P. F., Cummings, M. L., Wilson, M. A., Long, E. G., Hurwitz, E. S., and Juranek, D. D., 1989, Large community outbreak of cryptosporidiosis due to contamination of a filtered public water supply, *N. Engl. J. Med.,* 320, 1372.

221. Rush, B. A., Chapman, P. A., and Ineson, R. W., 1990, A probable waterborne outbreak of cryptosporidiosis in the Sheffield area, *J. Med. Microbiol.,* 32, 239.
222. Smith, H. V. and Smith, P. G., 1990, Parasitic protozoa in drinking water, *Endeavour,* 14, 74.
223. Mackenzie, W. R., Hoxie, N. J., Proctor, M. E., Gradus, M. S., Blair, K. A., Peterson, D. E., Kazmierczak, J. J., Addiss, D. G., Fox, K. R., Rose, J. B., and Davis, J. P., 1994, A massive outbreak in Milwaukee of Cryptospridium infection transmitted through the public water supply, *New Eng. J. Med.*, 331, 161.
224. Kramer, M. H., Herwaldt, N. J., Craun, G. F., Calderon, R. L., and Juranek, D. D., 1996, Surveillance for waterborne disease outbreaks, United States, 1993–1994, *J. Am. Water Works Assoc.,* 88, 66.
225. Smith, H. V. and Rose, J. B., 1998, Waterborne cryptosporidiosis: current status, *Pasasitol. Today,* 14(1), 14.
226. Badenoch, J., 1990, Cryptosporidiosis—a waterborne hazard (opinion), *Lett. Appl. Microbiol.*, 11, 269.
227. Ainsworth, R. G., 1990, Water treatment for health hazards, *J. Inst. Water Environ. Manage.,* 4, 489.
228. Barer, M. R. and Wright, A. E., 1990, *Cryptosporidium* and water: a review, *Lett. Appl. Microbiol.,* 11, 271.
229. Juranek D. D., 1995, Cryptosporidiosis: source of infection and guidelines for prevention, *Clin. Infect. Dis.,* 21, S57.
230. Department of the Environment, 1995, Drinking water 1994, a report of the chief inspector, Drinking Water Inspectorate, HMSO, London.
231. Department of the Environment, Department of Health, 1990, Cryptosporidium in water supplies, report of a group of experts, HMSO, London.
232. **Anon.,** 1996, *Cryptosporidiosis* in England and Wales, *Commun. Dis. Rep.*, 6, 20.
233. Rogers, J. and Keevil, C. W., 1995, Survival of *Cryptosporidium parvum* oocysts in biofilm and planktonic samples in a model system, in *Protozoan Parasites and Water,* Betts, W. B., Casemores, D., Fricker, C., Smith, H., and Watkins, J., Eds., The Royal Society of Chemistry.
234. Rogers, J. and Keevil, C. W., 1995, Survival of *Cryptosporidium parvum* in aquatic biofilm, in *Protozoal Parasites in Water,* Thompson, C. and Fricker, C., Eds., Royal Society of Chemistry, London, 209.
235. Akin, E. W. and Jakubowski, P., 1986, Drinking water transmission of *Giardiasis* in the United States, *Water Sci. Technol.,* 18, 219.
236. LeChevallier, M. W., Norton, W. D., and Lee, R. G., 1991, Occurrence of *Giardia* and *Cryptosporidium* spp. in surface water supplies, *Appl. Environ. Microbiol.,* 57, 2610.
237. LeChevallier, M. W., Norton, W. D., and Lee, R. G., 1991, *Giardia* and *Cryptosporidium* spp. in filtered drinking water supplies, *Appl. Environ. Microbiol.,* 57, 2617.
238. Browning, J. R. and Ives, D. G., 1987, Environmental health and the water distribution system: a case history of an outbreak of giardiasis, *J. Inst. Water Environ. Manage.,* 1, 55.
239. Jephcote, A. E., Begg, N., and Baker, I., 1986, An outbreak of giardiasis associated with mains water in the United Kingdom, *Lancet,* 8483, 730.
240. Craun, G. F., 1977, Waterborne outbreaks, *Water Pollut. Cont. Federat.,* 49, 1268.
241. Lin, S. D., 1985, *Giardia lambia* and water supply, *J. Am. Water Works Assoc.,* 77, 40.
242. West, P. A., 1991, Human pathogenic viruses and parasites: emerging pathogens, *J. Appl. Bacteriol.,* Symposium Supplement, 70, 1075.
243. Craun, G. F., 1986, *Waterbourne Diseases in the United States,* CRC Press, Boca Raton, FL.

244. Buisson, Y., Vancuyckgandre, H., and Deloince, R., 1993, Water and viral hepatitis, *Bull. Soc. Pathol. Exot.,* 86(5), 479.
245. American Public Health Association, 1992, *Standard Methods for the Examination of Water and Wastewater,* 18th ed., American Public Health Association, Washington, D.C.
246. Geldenhuys, J. C. and Pretorius, P. D., 1989, The occurrence of enteric viruses in polluted water, correlation to indicator organisms and factors influencing their numbers, *Water Sci. Technol.,* 21, 105.
247. Bitton, G., 1978, Survival of enteric viruses, in *Water Pollution Microbiology,* Vol. 2, Mitchell, R., Ed., Wiley Interscience, New York, 273.
248. Sellwood, J. and Dadswell, J., 1991, Human viruses and water, in *Current Topics in Clinical Virology,* Morgan-Caprier, P., Ed., Laverham Press, Salisbury, 29.
249. Craun, G. F., 1991, Cause of waterbourne outbreaks in the United States, *Water Sci. Technol.,* 24, 17.
250. Crabtree, K. D., Gerba, C. P., Rose, J. B., and Haas, C. N., 1997, Waterborne adenovirus: a risk assessment, *Water Sci. Technol.,* 35, 1.
251. vanden Bossche, G. and Krietermeyer, S., 1994, Detergent conditioning of biofilm samples: a most sensitive method for the detection of enterovirus infectivity, paper presented to the IAWQ Health-Related Water Microbiology Symposium, Budapest.
252. Quignon, F., Kiene, L., Levi, Y., Sardin, M., and Schwartzbrod, L., 1997, Virus behaviour within a distribution system, *Water Sci. Technol.,* 35, 311.
253. Levy, R. V., 1985, Invertebrate protection of coliforms and heterotrophic bacteria, *Proc. Seminar on Current Status of Drinking Water Microbiology,* AWWA Annual Conference, Washington, D.C.
254. Payment, P., Richardson, L., and Siemiatychi, J., 1991, Gastrointestinal health effects associated with the consumption of drinking water produced by point of use domestic reverse osmosis filtration units, *Appl. Environ. Microbiol.,* 57, 945.
255. Surman, S., Lee, J. V., Rowe, T., and Nichols, G., 1998, The role of toxins and virulence factors in heterotrophic bacteria present in drinking water supplies, *J. Appl. Microbiol.,* Symposium Supplement, 84, 150S.

9 Methods of Sampling Biofilms in Potable Water

CONTENTS

9.1 INTRODUCTION

On a routine basis water sampled from potable water systems is assessed for total viable numbers and, depending on the situation, specific bacteria, protozoa and fungi. These analyses provide results which can be interpreted and compared time after time across a range of samples and environments and are understood by microbiologists and a wide range of water treatment specialists.

Sampling for biofilms on a routine basis is more complex and uses two main types of microbial analysis

1. Noninvasive which leaves the substrata intact, such as swabbing.
2. Invasive which breaks the integrity of the system, such as sampling pipe sections.

The latter invasive techniques interrupt water flow and can potentially lead to contamination of an otherwise uncontaminated water system such as those that supply dialysis units or purified water systems. Other problems that have to be considered are the pressure drop leading to sloughing off of the biofilm into the water phase and subsequent blocking of filters.

9.2 NONINVASIVE SAMPLING METHODS

Swabbing of surfaces provides a noninvasive method (to the pipe system) of sampling for microbial biofilms attached to surfaces. This technique can be used to qualitatively assess for the presence of the total viable count and specifically for microorganisms such as *Legionella pneumophila*.[1-3] However, the biofilm structure, itself, would be destroyed when removed from the surface using swabbing. The technique incorporates removing a sterile swab from its sheath and swabbing surfaces of, for example, a shower head, inside a faucet or valve system, or inside the wall of a water tank. The swab would then be broken off into an aqueous phase of stabilising fluid such as Pages Amoebal Saline (PAS) and kept at less than 10°C until laboratory analysis. The sample could then be processed using a number of serial dilutions and various growth media to determine the presence or absence of microorganisms including fungi.

In domestic cold water tanks, biofilm formation can be monitored by placing sections of similar material, the same composition as the tank, such as glass reinforced plastic (GRP) into the water phase. The coupon material can be suspended from inert wire such as titanium or a plastic that will be known not to leach microbial nutrients into the water phase which could interfere with the experiment. Such techniques have been used by Pavey et al.[4,5] and Walker et al.[6] to carry out time course experiments within a full scale domestic hot and cold water system to determine colonisation of *L. pneumophila* in biofilms before and after disinfection studies. Similar techniques have been used to investigate biofouling control strategies within calorifiers, evaporative condensers, and cooling tower ponds.

9.3 INVASIVE SAMPLING TECHNIQUES

9.3.1 Directly Implanted Devices

These devices are designed to be used in high pressure fittings and include the Cosaco 2 in. access fitting assembly (Figure 9.1). The benefit of the Cosaco device is that it can be serviced at the pressure within the system and thus avoids the

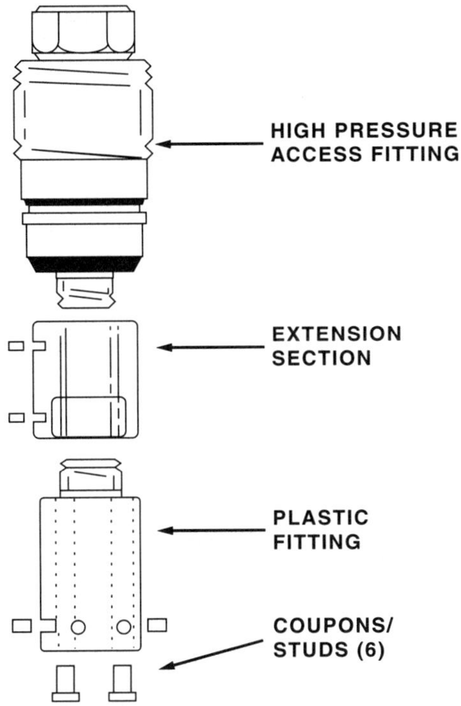

FIGURE 9.1 Cosaco high pressure fitting assembly.

requirement for a partial shut down, depressurisation, and drainage. The coupons have an exposed surface area of 0.5 cm^2 and any material can be used to determine fouling capabilities or monitor biofouling on the material the system is made out of over a time course.[7]

9.3.2 Side Stream Devices—Robbins Device

These devices have been developed primarily for the oil industry but have served as the basis for cross over into hot and cold water systems (Figure 9.2) The most commonly known device is the Robbins Device.[8,9] The original device was made from admiralty brass, but the devices can be commissioned in any material of choice to suit a particular water system. These systems have sample points at 12, 3, 6, and 9 o'clock along the length of the pipe and are put in place parallel to the original pipework. The increased number of sampling points allow for long term monitoring of biofouling of materials to determine suitability in such systems and for disinfection efficacy testing which needs to be monitored. Replicate analysis can also be ascertained to validate any statistical analysis. Corrosion monitoring can also be carried out on the removed coupons. Parallel to the main system, the Robbins device can be valved off to allow removal of coupons without affecting the main flow rate. It must be borne in mind that such parallel devices generally only receive a portion of the flow and as such have to be engineered to achieve flow rates similar to the

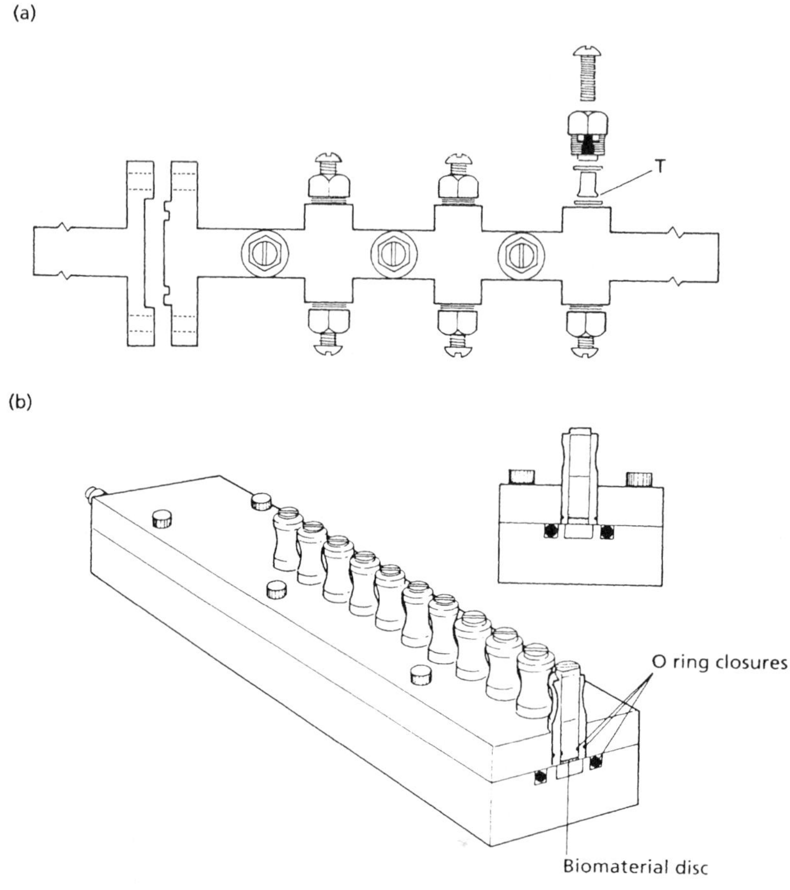

FIGURE 9.2 (a) Diagram of a Robbins device biofilm sampler [removable test surfaces (T) have a 0.5 cm^2 area exposed to fluid]; (b) cross-sectional view of a modified Robbins device. Courtesy of Hopton, J.W. and Hill, E.C., Eds., *Industrial Microbiological Testing*, 1987, with permission from Blackwell Scientific.

main piping (approximately 1.3 m per s). However, this decreased flow rate can be used to simulate the worst case scenario of flow rate and, hence, the greatest opportunity for biofouling. Similar types of devices were used at the county hospital in Hellersen, Germany to investigate corrosion and biofilm formation.[10]

9.3.3 Distribution System Test Rig

As far as potable water is concerned, there is also the problem of biofilms in the distribution network.[11] A number of large scale test rigs have been engineered to replicate distribution networks. One of these is the pipe rig at Kempton Park Advanced Water Treatment Centre, the world's longest experimental one pass pipe distribution system[12] [Figures 9.3(a) and 9.3(b)]. The 1.3. km pipe has a 110 mm diameter pipe distribution system and is fed with surface derived water. The water

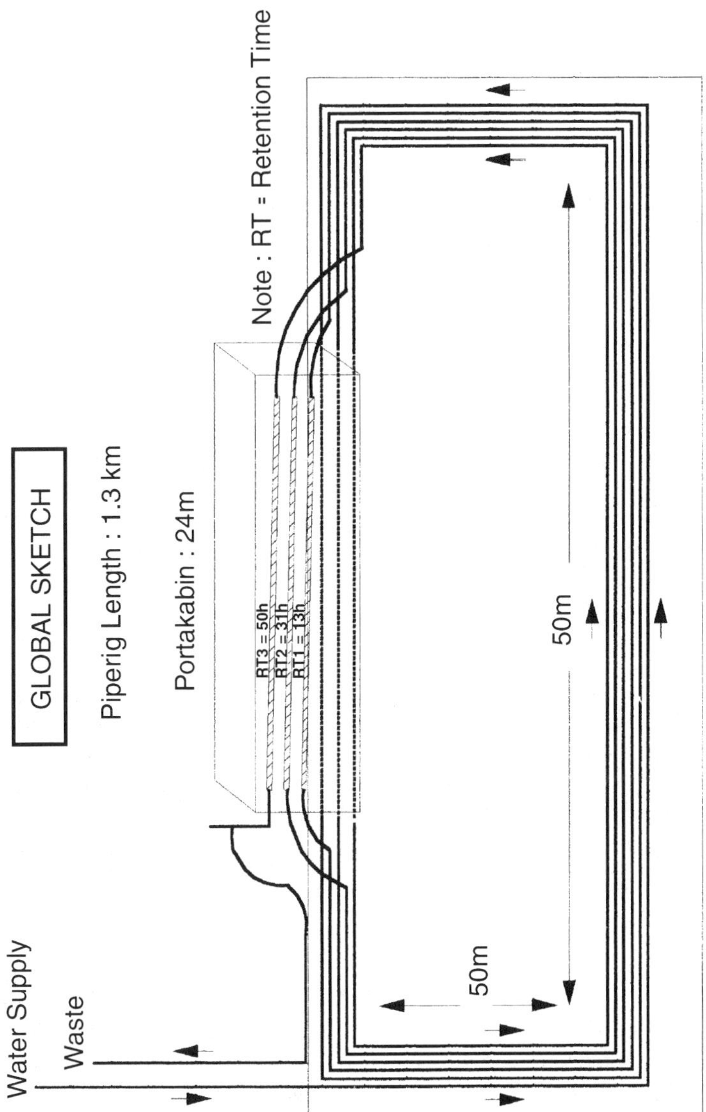

FIGURE 9.3a Global sketch of piperig. Courtesy of Thames Water Utilities.

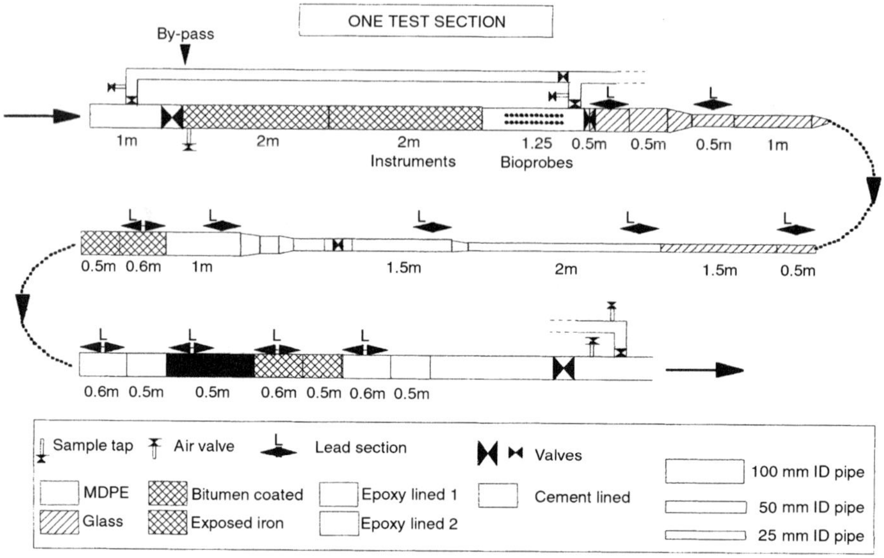

FIGURE 9.3b Global sketch of the pipe test rig. Courtesy of Thames Water Utilities.

passes through a balancing tank, then goes underground, and has three sections that pass through an experimental portacabin at 0.5, 0.9, and 1.3 km along the length of the pipe where biofilm and water samples are taken.

9.3.4 NCIMB Biofilm Generator

This is a cheap and easy way to build a biofilm generator made from high density polyethylene pipe (Figure 9.4).[13] A flat area is then planed along the surfaces, approximately 10 mm wide. At uniform distances along the length of the pipe, holes are drilled. T-shaped coupons or studs fitted with o-rings are pressed into the holes and held down using unex hose clips. Although only previously used within a laboratory for studies on mild steel induced biocorrosion, this system could be used for domestic biofilm studies under controlled conditions.

9.3.5 Biotube

This is a flow through drainable tube section that has removable and replaceable biofouling surfaces [Figures 9.5(a) and (b)]. The section is placed parallel to the main flow with all material to be tested attached to one removable section. These bypass rigs are supplied by Metal Supplies Limited and can be distributed in multiples such that a long term monitoring programme can be put in place, if necessary. A number of the large water treatment companies such as Aquazur (formally known as Houseman's) [Figures 9.6(a) and (b)] also supplies and fits similar biomonitors for hot and cold water systems. These systems can be specifically designed for each system and, preferentially, should be installed as the system is being built. Materials

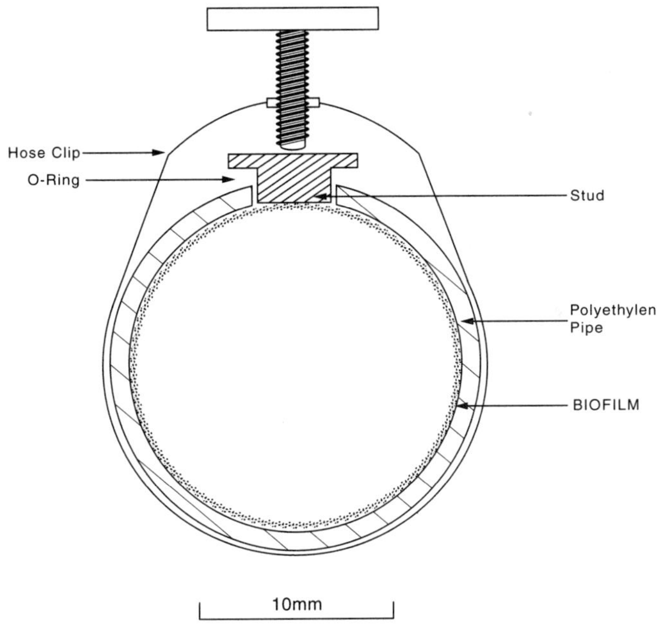

FIGURE 9.4 Cross section of the pipe of the NCIMB biofilm generator. Courtesy of Hopton, J.W. and Hill, E.C., Eds., *Industrial Microbiological Testing,* 1987, with permission from Blackwell Scientific.

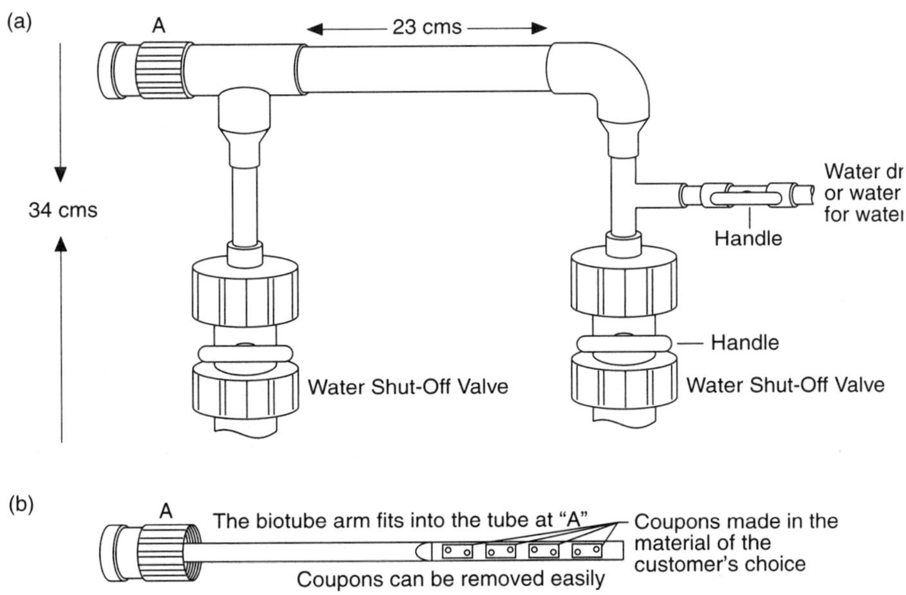

FIGURE 9.5 (a) Schematic of biotube; (b) the biotube arm which fits into the biotube at A.

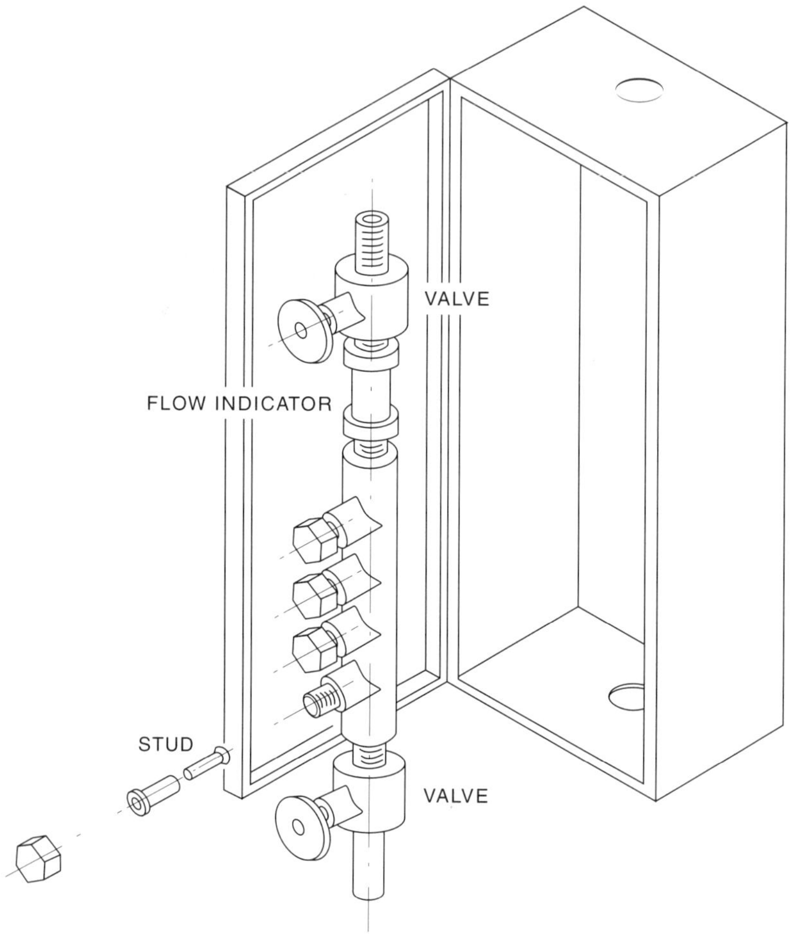

FIGURE 9.6a Biofilm monitor.

chosen depend on suitability and specification from a choice including PTFE, PVDF, and stainless steel.

9.3.6 Pipe Replacement and Trombone Systems

In some instances such as in pipe failure where a section of tubing has to be replaced, the removed pipe can be used for analysis. In other cases, pipe section removal is the only way to investigate biofouling when monitoring had not been thought of and no monitoring devices are available. A trombone arrangement of removable pipework was engineered by Pavey et al.[4] where sections could be routinely removed and the trombone reattached to allow flow. This enabled the actual pipework in use to be analysed for biofilm development.

Upon removal, the pipe section needs to be kept hydrated and end-capped, filled with source water, sealed with another end-cap, and refrigerated until analysis.

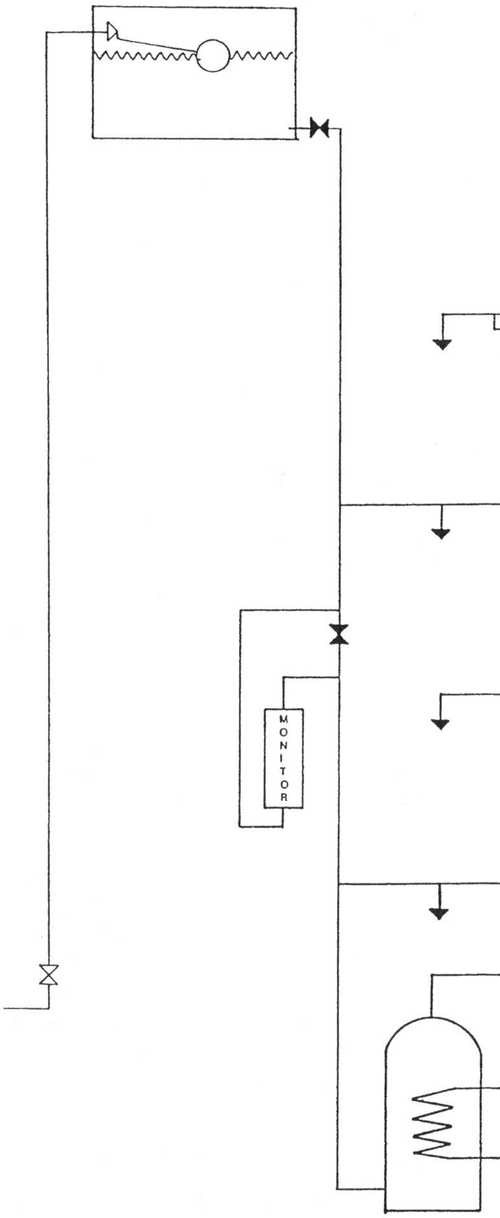

FIGURE 9.6b Possible position for the installation of biofilm monitors on a hot/cold water system.

9.3.7 Taps and Valves

Although invasive and sacrificial, the destruction of tap, shower, and valve fittings provide a method of obtaining material for biofilm analysis. Such techniques were again used by Pavey et al.[4] to determine if their disinfection strategies of chlorine dioxide and ionisation had controlled *L. pneumophila* biofilms on washers within the fittings. Interestingly, the results indicated that *L. pneumophila*, attached to the tap and shower washers, was shown to have survived the disinfection process and would have acted as a seed for when the system was used next.

9.4 DESTRUCTIVE TECHNIQUES FOR BIOFILM ANALYSIS

9.4.1 Direct Surface Plating

The most common laboratory method of analysis of surface-associated bacteria is to remove the bacteria, vortex to disperse the microorganisms, and then use plate count assays to asses the numbers and type of microbial flora present.[14] These techniques are inherently variable. For example, the bacteria have to be removed from the surfaces, and such methods can include

- Scraping with a sterile dental probe to physically remove the bacteria from the surface.[15]
- Sonticating thrice in 10- or 20-second bursts.
- Vortexing for 1 min and repeating after 5 min on ice.[16]
- Vortexing with glass beads.

Following dispersion, serial dilutions can be carried out to determine the numbers and species variation of the bacteria present. From personal experience, physical methodologies are required to actually remove a biofilm from a surface, not just shaking/vortexing the sample and hoping that the microorganisms have been removed.

9.4.2 Contact Plating

This technique is primarily used in the food industry to detect contamination on food hygiene surfaces. However, Eginton et al.[17] described a contact plating method that reproducibly quantifies the ease of removal of microorganisms from surfaces. Colonised test surfaces were repeatedly transferred through a succession of agar plates providing a measure of the strength of attachment to a surface.

9.4.3 ATP Analysis

All living microbial cells contain adenosine triphosphate (ATP) which may be extracted and subsequently assayed with the enzyme firefly luciferase. The amount of light generated by this enzymic reaction can be measured in a suitable luminometer and is directly related to the ATP extracted and, thus, the number of microbial cells present. The measurement of ATP extracted from adhered cells[18] offers the advantages of being both rapid and simple to perform. Blackburn et al.[19] described a rapid technique for studying microbial adhesion to polyethylene and stainless steel using *Pseudomonas* spp. and compared the results to direct microscopical counts demonstrating a good correlation between the ATP assay and the counts. However, its use for quantifying biofilms from water systems may be more complex, owing to environmental products such as irons and humic and fulvic acids interfering with the reaction, and thus care must be taken in the interpretation of results. With ATP analysis, one obtains a total assessment of the ATP from all the cells present, and it is not discriminatory to only bacterial ATP.

9.4.4 Fatty Acid Analysis

Lipopolysaccharides (LPS) are cellular constituents of almost all Gram-negative bacteria. Because they contain unique carbohydrates and fatty acids, LPS has been shown to be a useful diagnostic marker for particular bacteria.[20] Sonesson et al.[21] described a unique structural type of LPS from *Legionella pneumophila,* and used gas chromatography–mass spectrometry analysis to detect *L. pneumohila* in potable water biofilms from copper and polybutylene materials.[22]

9.5 NONDESTRUCTIVE TECHNIQUES FOR BIOFILM ANALYSIS

9.5.1 Microscopy

Microscopy has provided a vast range of methodologies with which to visualise biofilms since Henrici[23] first made detailed morphological observations of attached bacteria and Staley[24] first determined the *in situ* growth rates of attached cells using an immersed microscope. This is important in biofilm assessment. Whereas viable recovery only assesses what has been able to be grown, the lack of detection of viable cells does not necessarily mean a viable (nonculturable or otherwise) biofilm is not attached to the surface. Even in the situation where the biofilm is not viable, the removal of a surface attached film can be beneficial to prevent corrosion cells, reduced flow, reduced heat transfer coefficient, and microbial cells reattaching easily to the surface.

9.5.2 Light Microscopy

On light transmittable surfaces such as glass, biofilm development can be assessed. However, discrimination can often be poor depending on the purity of the water or medium used to grow the biofilms. In 1978, Zimmermann, Iturriaga, and Becker-Birck[25] discussed a technique to use the microbial universal electron transport system to reduce 2-(*p*-iodophenyl)-3-(*p*-nitrophenyl)-5-phenyl tetrazolium chloride (INT) to INT-formazon. Respiring bacterial cells deposit INT-formazon as optically dense dark red intracellular spots and after enumeration with phase contrast, numbers of respiring bacteria were determined. Lawrence, Korber, and Caldwell[26] used dark field and image analysis to quantify colonisation kinetics, growth rates, and interactions of microorganisms on surfaces.

9.5.3 Hoffman Modulation

Hoffman modulation contrast microscopy is an adaptation of bright field microscopy and allows noninvasive imaging without the need for prior staining. Biofilms can be imaged as a three-dimensional structure with high contrast resolution. The three-dimensional image effect was obtained by conversion of opposite phase gradients to opposite intensities in the image.[27] This type of microscopy has been used to study the presence of amoeba grazing on biofilms.[28] For this microscopy to be useful, the biofilm has to be adhered to glass (or equivalent see-through) substrata.

9.5.4 Differential Interference Contrast

Lambe, Fergusen, and Ferguson[29] used differential interference contrast (DIC) in the conventionally transmitted form to view the glycocalyx of *Bacteroides* on glass coverslips. However, Keevil and Walker[30] used an adapted DIC microscope to view biofilms from above powered through a mercury lamp, allowing biofilms on optically dense surfaces to be viewed. Adaptations included siting of the polariser above the specimens, allowing opaque specimens to be viewed, appropriate filter blocks, and the presence of mirror plates in the mercury lamp housing to increase the light intensity.

9.5.5 Fluoresence

Many fluorescent stains have been adapted to visualise biofilms on surfaces including INT,[25] acridine orange[31] (AD), propidium iodide,[32] rhodamine (Rh 123),[33] and 5-cyano-2,3-ditoyl tetrazolium chloride (CTC).[34]

The fluoresence colours obtained with AO are to some extent a reflection of the viability of the cells. Slow growing cells have a tightly coiled DNA and a low RNA content, and thus the cells fluoresce in the green part of the spectrum owing to the interchelation of the AO with nucleic acid. By contrast, dead cells have loosely coiled DNA whilst rapidly growing cells have a high RNA content—cells in either of these states fluoresce in the yellow/orange part of the spectrum. The use of AO for viability assays is immensely difficult to interpret when working with environmental samples owing to different growth rates and species diversification resulting in a wide range of cellular colour. In addition to this, AO often interacts with foreign bodies making differentiation of microbial cells difficult. In some samples, propidium iodide (a structural probe for alternating and nonalternating DNA polymers containing guanine and binding to DNA by intercalation)[35] is used as an alternative. Fluorescence microscopy coupled to image analysis provides a relatively rapid technique for percentage coverage assessment of biofouling and is currently used in industry.

9.5.6 Scanning Confocal Laser Microscopy (SCLM)

Scanning electron microscopy allows noninvasive visualisation of complex structures at high magnification and it can be applied where debris obscures the image.

Coupled with on-line computer image enhancement, SCLM can be used to visualise and quantitate biofilm growth and structure. There have been a number of researchers whose work has pioneered the use of SCLM to study attached bacteria, such as Caldwell and Lawrence.[36] SCLM allows detailed visualisation of thick microbiological samples in cases in which the applications of traditional microscopy are limited, such as allowing out-of-focus haze, horizontal and vertical sectioning, determination of three-dimensional relationships, and three-dimensional reconstructions from optical thin sections.[37] Recent applications of SCLM using immunofluorescent labelling have shown that biofilms can lead to substrata corrosion,[38] have a complex structure consisting of cell–cell interactions,[39-41] form discrete aggregates of microbial cells in an EPS matrix identified using green fluorescent protein,[42] possess interstitial voids and open channels, and are shown using fluorescein and phycoerythrin.[43]

9.5.7 Atomic Force Microscopy (AFM)

AFM is a form of scanning probe microscopy that uses a sharp probe to map the contours of a sample and generates surface profiles of cellular structures at the atomic level.[44] It is a type of near field microscope and is, therefore, not limited in its resolution by diffraction effects.[45]

9.5.8 Scanning Electron Microscopy (SEM)

SEM allows the visualisation of complex structures at high magnification. Typically, materials are fixed in osmium tetroxide, then dehydrated in a series of alcohol/water solutions such as 30, 50, 70, and 90% before being gold sputtered and viewed as secondary electron images. However, shrinkage and loss of samples have occurred through sample preparation.[46,47] The consequences are that the subsequent images may not represent the actual shape, morphology, and structure of the biofilm that was on the surface.

9.5.9 Transmission Electron Microscopy (TEM)

TEM provides internal cross-sectional detail of the individual microorganisms and their relationship to each other. The biofilms are typically fixed in gluteraldehyde/ruthenium red in cacodylate buffer before being fixed in osmium tetroxide/ruthenium red. The sample is then placed through a series of graded alcohols before being placed in propylene oxide and spurr resin from which ultra thin sections are prepared. This technique has been used to good effect to demonstrate the spatial differentiation between cells filled by a matrix of polymers stained by the ruthenium red[48] as well as positioning of cells within the biofilm as determined by aerobic/anaerobic conditions.[49]

9.5.10 Environmental SEM (ESEM)

The problems associated with distortion during preparation using SEM and TEM can be alleviated using ESEM. With this technique the chamber is differentially pumped, allowing it to operated with up to 10 torr of water vapour and, thus, enabling the specimens to be viewed in their true hydrated state. Such processing is a major advance over the time for preparation and surface morphology distortion that occurs with conventional EM.

9.6 ACKNOWLEDGMENTS

We would like to thank Sarah McMath of Thames Water for supplying diagrams of the Kempton 1.3 KM distribution test rig and Jim Robinson of Aquazur for sending photos of their potable water biofilm monitor.

9.7 REFERENCES

1. Wadowsky, R. M., Yee, R. Y., Mezmar, L., Wingadn, E. J., and Dowlilng, J. N., 1982, Hot water systems as sources of *Legionella pneumophila, Arch. Microbiol.,* 153, 72.
2. Colbourne, J. S., Pratt, D. J., Smith, M. G., Fisher-Hoch, S. P., and Harper, D., 1984, Water fittings as sources of *Legionella pneumophila* in a hospital plumbing system, *Lancet,* 1, 210.
3. Walker, J. T., Mackerness, C. W., Mallon, D., Makin, T., Williets, T., and Keevil, C. W., 1995, Control of *Legionella pneumophila* in a hospital water system by chlorine dioxide, *J. Ind. Microbiol.,* 15, 384.
4. Pavey, N. L., Walker, J. T., Ives, S., Morales, M., and West, A. A., 1996, *Ionisation Water Treatment—for Hot and Cold Water Services,* Technical Note TN 6/96, Bourne Press.
5. Pavey, N. L., Roper, M., Walker, J. T., Lucas, V., and Roberts, A. D. G., 1998, *Chlorine Dioxide Water Treatment—for Hot and Cold Water Services,* Technical Note TN, Bourne Press.
6. Walker, J. T., Morales, M., Ives, S., and West, A. A., 1997, Controlling *Legionella* and biofouling using silver and copper ions: fact or fiction, in *Biofilms—Communities and Interactions,* Bioline, Cardiff, 279.
7. Gilbert, P. D. and Herbert, B. N., 1987, Monitoring microbial fouling in flowing systems using coupons, in *Industrial Microbiological Testing,* Hopton, J. W. and Hill, E. C., Eds., Technical Series 23, Society for Applied Bacteriology, Blackwell Scientific, Oxford, 79.
8. McCoy, W. F., Bryers, J. D., Robbins, J., and Costerton, J. W., 1981, Observations of fouling biofilm formation, *Can. J. Microbiol.*, 27(9), 910.
9. McCoy, W. F. and Costerton, J. W., 1982, Fouling biofilm development in tubular flow systems, *Dev. Ind. Microbiol.*, 23, 551.
10. Wagner, D., Fischer, W., and Paradies, H. H., 1992, Copper deterioration in a water distribution system of a county hospital in Germany caused by microbial influenced corrosion—simulation of the corrosion process in two test rigs installed in this hospital, *Werst. Und. Korr.,* 43, 496.
11. LeChevallier, M. W., Babcock, T. M., and Lee, R. G., 1987, Examination and characterisation of distribution systems biofilms, *Appl. Environ. Microbiol.,* 53(12), 2714.
12. McMath, S. M., Sumpter, C., Holt, D. M., Delanoue, A., and Chamberlain, A. H. L., 1999, The fate of environmental coliforms in a model water distribution system, *Lett. Appl. Microbiol.,* 28, 93.

13. Green P. N., Bousfield, I. J., and Stones, A., 1987, The laboratory generation of biofilms and their use in biocide evaluation, in *Industrial Microbiological Testing,* Hopton, J. W. and Hill, E. C., Eds., Technical Series 23, Society for Applied Bacteriology, Blackwell Scientific, Oxford, 99.
14. Rogers, J., Dowsett, A. B., Dennis, P. J., Lee, J. V., and Keevil, C. W., 1994, Influence of temperature and plumbing material selection on biofilm formation and growth of *Legionella pneumophila* in a model potable water containing complex microbial flora, *Appl. Environ. Microbiol.,* 60, 1585.
15. Brading, M. G., Boyle, J., and Lappin-Scott, H. M., 1995, Biofilm formation in laminar flow using *Pseudomonas fluorecesns* EX101, *J. Ind. Microbiol.,* 15, 297.
16. Linton, C. J., Sherriff, A., and Millar, M. R., 1999, Use of a modified Robbins device to directly compare the adhesion of *Staphylococcus epidermidis* RP62A to surfaces, *J. Appl. Microbiol.,* 86, 194.
17. Eginton, P. J., Gibson, H., Holah, J., Handley, P. S., and Gilbert, P., 1999, Strength for adhesion of bacteria to surfaces in biofilms, in *The Life and Death of Biofilm,* Wimpenny, J., Handly, P., Gilbert, P., and Lappin-Scott, H., Eds., Bioline, Cardiff, 61.
18. Harber, M. J., MacKenzie, R., and Asscher, A. W., 1983, A rapid bioluminescence method for quantifying bacterial adhesion to polystyrene, *J. Gen. Microbiol.,* 129, 621.
19. Blackburn, C. W., Gibbs, P. A., Roller, S. D., and Johal, S., 1989, Use of ATP in microbial adhesion studies, in *ATP Luminescence—Rapid Methods in Microbiology,* Stanley, P. E., McCarthy, B. J., and Smither, R., Eds., Technical Series 23, Society for Applied Bacteriology, Blackwell Scientific, Oxford, 145.
20. Jantzen, E. and Bryn, K., 1985, Whole-cell and lipopolysaccharide fatty acids and sugars of gram negative bacteria, in *Chemical Methods in Bacterial Systematics,* Goodfellow, M. and Minnikin, D. E., Eds., Technical Series 20, Society for Applied Bacteriology, Academic Press, London, 145.
21. Sonesson, A., Jantzen, E., Bryn, K., Larsson, L., and Eng, J., 1989, Chemical composition of a lipopolysaccharide from *Legionella pneumophila, Arch. Microbiol.,* 153, 72.
22. Walker, J. T., Sonesson, A., Keevil, C. W., and White, D. C., 1993, Detection of *Legionella pneumophila* in biofilms containing a complex microbial consortium by gas chromatography–mass spectrometry analysis of genus specific hydroxy fatty acids, *FEMS Microbiol. Lett.,* 113, 139.
23. Henrici, A. T., 1932, *J. Bacteriol.,* 25, 277.
24. Staley, J. T., 1971, Growth rates of algae determined *in situ* using an immersed microscope, *J. Phycol.,* 7, 13.
25. Zimmermann, R., Iturriaga, R., and Becker-Birck, J., 1978, Simultaneous determination of the total number of aquatic bacteria and the number thereof involved in respiration, *Appl. Environ. Microbiol.,* 36, 926.
26. Lawrence, J. R., Korber, D. R., and Caldwell, D. E., 1989, Computer enhanced darkfield microscopy for the quantitative analysis of bacterial growth and behaviour on surfaces, *J. Microbiol. Meth.,* 10, 123.
27. Hoffman, R., 1977 The modulation contrast microscope, *J. Microsc.,* 110, 205.
28. Surman, S. B., Walker, J. T., Goddard, D. T., Morton, L. H. G., Keevil, C. W., Weaver, W., Skinner, A., and Kurtz, J., 1996, Comparison of microscope techniques for the examination of biofilms, *J. Microbiol. Meth.,* 25, 57.
29. Lambe, D. W., Ferguson, K. P., and Ferguson, D. A., 1988, The *Bacteroides* glycocalyx as visualised by differential interference contrast microscopy, *Can. J. Microbiol.,* 34, 1189.
30. Keevil, C. W. and Walker, J. T., 1992, Nomarski DIC microscopy and image analysis of biofilms, *Binary,* 4, 92.

31. Daley, R. J. and Hobbie, J. E., 1975, Direct counts of aquatic bacteria by a modified epifluorescent technique, *Limnol. Oceanogr.,* 20, 875.
32. Jones, K. H. and Senft, J. A., 1985, An improved method to determine cell viability by simultaneous staining with fluorescence diacetate-propidium iodide, *J. Histochem. Cythochem.,* 33, 77.
33. Kaprelyants, A. S. and Kell, D. B., 1992, Rapid assessment of bacterial viability and vitality by rhodamine 123 and flow cytometry, *J. Appl. Bacteriol.,* 72, 410.
34. Rodriguez, G. G., Phipps, D., Ishiguro, K., and Ridgway, H. F., 1992, Use of a fluorescent redox probe for direct visualisation of actively respiring bacteria, *Appl. Environ. Microbiol.,* 58, 1801.
35. Wilson, W. D., Wang, Y., Krishnamoorthy, C. R., and Smith, J. C., 1986, Intercalators as probes of DNA conformation: propidium binding to alternating and non-alternating polymers containing guanine, *Chem. Biol. Interact.,* 58, 41.
36. Caldwell, D. E. and Lawrence, J. R., 1988, Study of attached cells to continous-flow slide culture, in *A Handbook of a Laboratory Model System for Microbial Ecosystem Research,* Wimpenny, W. T., Ed., CRC Press, Boca Raton, FL, 117.
37. Lawrence, J. R., Korber, D. R., Hoyle, B. D., Costerton, J. W., and Caldwell, D. E., 1991, Optical sectioning of microbial biofilms, *J. Bacteriol.,* 173, 6558.
38. Walker, J. T., Hanson, K., Caldwell, D., and Keevil, C. W., 1998, Scanning confocal laser microscopy study of biofilm induced corrosion on copper plumbing tubes, *Biofouling,* 12, 333.
39. Cummings, D., Moss, M. C., Jones, C. L., Howard, C. V., and Cummins, P. G., 1992, Confocal microscopy on dental plaques development, *Binary,* 4, 86.
40. Doolittle, M. M., Cooney, J. J., and Caldwell, D. E., 1996, Tracing the interaction of bacteriophage with bacterial biofilms using fluorescent and chromogenic probes, *J. Ind. Microbiol.,* 16(6), 331.
41. Sanford, B.A., de Feijter, A. W., Waden, M. H., and Thomas, V. L., 1996, A dual fluorescence technique for visualization of *Staphylococcus epidermidis* biofilm using scanning confocal laser microscopy, *J. Ind. Microbiol.,* 16(1), 48.
42. Kuehn, M., Hausner, M., Bungartz, H. J., Wagner, M., Wilderer, P. A., and Wuertz, S., 1998, Automated confocal laser scanning microscopy and semiautomated image processing for analysis of biofilms, *Appl. Environ. Microbiol.*, 64, 4115.
43. De Beer, D., Srinivasan, R., and Stewart, P. S., 1994, Direct measurement of chlorine penetration into biofilms during disinfection, *Appl. Environ. Microbiol.,* 60, 4339.
44. Goddard, D. T., 1993, Imaging soft and delicate materials, *Mat. World,* 1, 616.
45. Hyde, F. W., Alberg, M., and Smith, K., 1997, Comparison of fluorinated polymers against stainless steel, glass and polypropylene in microbial biofilm adherence and removal, *J. Ind. Microbiol. Biotechnol.,* 19(2), 142.
46. Woldringh, C. L., de Jong, M. A., van den Berg, W., and Koppes, L., 1977, Morphological analysis of the division cycle of two *Escherichia coli* substrains during slow growth, *J. Bacteriol.,* 131, 270.
47. Chang, H. T. and Rittman, B. E., 1986, Biofilm loss during sample preparation for scanning electron micrscopy, *Water Res.,* 20, 1451.
48. Marrie, T. J. and Costerton, J. W., 1984, Scanning and transmission electron microscopy of *in situ* bacterial colonization of intravenous and intraarterial catheters, *J. Clin. Microbiol.,* 19, 687.
49. Kinniment, S. L., Wimpenny, J. W., Adams, D., and Marsh, P. D., 1996, Development of a steady-state oral microbial biofilm community using the constant-depth film fermenter, *Microbiology,* 142, 631.

10 Materials Used in the Transport of Potable Water with Special Reference to Stainless Steel and Corrosion

CONTENTS

10.1 INTRODUCTION

There are a large number of areas present in potable water systems where biofilm development can proliferate. These include pipes, valves, fire hydrants, gaskets, sealants, and lubricants. However, the emphasis in this chapter will be on the development of biofilms and the effects they have on pipe materials used in the transport of potable water.

Over the centuries there have been a large number of different materials used in the transport of potable water. Traditionally, these piping materials have included clay, wood, stone, and lead. Piping materials which are used today to carry potable water are numerous. These include ductile cast iron, cast iron lined with cement, steel, reinforced concrete, asbestos combined with portland cement, copper, and plastics, namely, polyvinyl chloride (PVC), polyethylene, polybutylene, and medium density polyethylene (MDPE).

The service life of all pipe materials is very important in potable water. This service life, however, is affected by a number of conditions including such things as water chemistry, microbial activity, climatic conditions, and corrosivity of the environment to which the pipe is exposed. In the case of cast iron and ductile iron pipe, the service life has been estimated at around 100 years; reinforced concrete, 50 years; and asbestos cement pipe; 30 years.[1] The service life of plastic pipes is uncertain because a large number of utilities is very reluctant to accept it as a replacement to the more traditional materials.

10.1.1 Materials and Biofilm Formation

Materials which are exposed to potable water are the major contributor to biofilm formation. Biofilm development will form on any pipe material which is exposed to potable water. The factors known to affect this process include the length of the pipe network, the predominant pipe material, age of the pipe, number of breaks per year, corrosion, sediment accumulation, and zones of static water.

A number of piping materials have led to many consumer complaints, particularly owing to the formation of corrosion products deposited at consumers' taps as a result of microbial activity. As far back as 1916, leather washers were found to cause deterioration in the bacterial quality of water owing to the colonisation of bacteria.[2]

10.2 PIPING MATERIALS USED IN THE SUPPLY OF POTABLE WATER

Within building service lines and pipe networks, the choice of material for transporting potable water is dependant upon a number of factors including cost, durability, appearance, climate conditions, biodegradability, strength, thermal properties (with respect to hot water systems), and workability (fittings and jointing methods).[3] Traditionally, four main materials have been used in commercial and municipal buildings for the supply of potable water.[4] These have included lead, galvanised steel, copper, and black/galavanized iron. Each of these piping materials will be considered in turn.

10.2.1 Lead

Lead is now prohibited as a new plumbing material in most European Community (EC) countries owing to the toxicity of the corrosion products formed which dissolve in the water (especially soft water) leading to concern about the public's health. Existing lead pipes are now being substituted with alternative materials. However, because of existing lead pipes which have not been replaced, the short term control measures being employed to reduce corrosion have involved the adjustments of pH in the potable water and the maintenance of central dosages of organophosphates.

10.2.2 Galvanised Steel

Generally, the use of zinc to coat steel pipes has provided a temporary measure in retarding corrosion and, thus, avoiding the formation of red water in stagnating plumbing systems. The good solubility of the zinc corrosion products (which are pH-dependant) and the uptake of zinc (and, possibly, lead and cadmium) into the water have led to a recommendation that galvanised steel pipes only be used in water with a pH greater than 7.3. Galvanised steels are widely used in hard water areas.

10.2.3 Copper

Copper pipes have suffered many problems, notably from pitting corrosion, particularly in cold, hard waters, resulting in what is described as type 1 corrosion and in soft acid hot water (greater than 60°C) by type 2 corrosion which has been shown to accelerate if traces of manganese are present in the water. Copper has been shown to corrode if water velocities have exceeded 1.2 ms^{-1}. Despite this, the maximum permitted water velocities that govern the suitability for copper usage in potable water fall from 4 ms^{-1} at 10°C to 2.5 ms^{-1} at 70°C.[5] Corrosion of copper has also been shown to accelerate if water pH falls below 7.[6] Manufacturers believe that chlorine levels of 1 to 2 mg per litre pose no problems to copper. It also has been documented that copper is known to not corrode when it is exposed to mg per litre chlorine doses of 20 to 50 for 1 to 3 hours. Extreme caution, however, is necessary when using copper, particularly in the presence of high levels of carbon dioxide as this is known to increase corrosion rates. This also has been evident when copper is exposed to high temperatures.[7]

10.2.4 Black/Galvanized Iron

Black and galanized iron has been used only occasionally in potable water.[8]

10.2.5 New and Emerging Materials Used in the Transport of Potable Water

The use of plastics, mainly PVC and MDPE, are beginning to replace traditional materials in potable water systems. However, polyvinyl plastics are known to leach chemicals and are very prone to sagging, particularly if water is left standing in them for long periods of time.[9]

As a legal requirement, materials used to transport potable water must not release chemicals at levels in excess of acceptable toxicity and health standards[10] (BS3505 and BS6920). Therefore, within hospitals and other municipal buildings, unplasticised PVC must be used in cold water systems only because they are known to leach toxic chemicals at high temperatures. The effect of chemicals leaching from plastics on human health is not known owing to a lack of documented research in this area.

In view of the continuing problem associated with piping materials,[11] particularly copper, stainless steel has been proposed as an alternative plumbing material for the supply of potable water in commercial buildings. However, there has been little documented research about its performance, effects on water quality, and susceptibility to biofilm development in potable water. One study has suggested that heavy metal ion leaching, specifically nickel, from stainless steel into potable water is evident, suggesting that this can be a temporary problem in newly plumbing piping systems.[12] However, this has since been rejected, provided that newly commissioned stainless steel piping is flushed through with potable water before any supply is made to the consumer. Any possible correlation between water quality and the different material qualities of stainless steel and the possible impact of increasing nickel allergies has not been researched at present. It is, however, unlikely that any correlation does exist as stainless steel has been found to be very stable in potable water, releasing only very low levels of metal ions into both potable water and biofilms.[13-15]

10.3 PROBLEMS ASSOCIATED WITH MATERIALS USED IN POTABLE WATER

10.3.1 Corrosion

Copper has long been the standard for many plumbing applications throughout the world,[16] with copper pipes representing 11% in Europe and 14% in the U.S. in 1989 of the total copper consumption.[17] However, problems encountered with copper piping when used for the plumbing of potable water systems in large institutional buildings (e.g., hospitals) have generated concern in various parts of the U.K.,[13,18,19] Germany,[20] Saudi Arabia,[21] and Japan. Problems noted are those of corrosion. The type of corrosion observed on copper takes the form of localised pitting or pinhole attacks evident on the inner surface of copper piping material which is generally confined to institutional buildings in soft water areas. Whilst pinhole corrosion does not result in catastrophic pipe failure, it does lead to severe shortening of a system's lifetime with disruption to the operation of municipal buildings, owing to increased incidences of repair work as a result of pipe failure and water leakage. Typical control regimes used in these systems included acid cleaning to disinfect the pipe material, filtration of the water to 0.2 μm, ultraviolet light treatment, and elevating temperatures to 60°C.[22]

Wagner, Fischer, and Paradies[23] have also described serious problems of copper corrosion in a county hospital in Germany, evident in cold and warm water systems, shortly after the hospital was opened. From this work, they established this so-called new type of corrosion characterised as

1. A biofilm consisting of polysaccharides.
2. Perforation of copper tubes in a short period of time.
3. Significant dissolution of copper into the potable water.[20]

Materials which were investigated as an alternative to copper as a result of premature corrosion in this German hospital, included stainless steel, cross-linked polyethylene (VPE), and polypropylene. It was found after 1 year that stainless steel piping used to carry potable water showed no evidence of general attack or pitting corrosion. A similar result was found in a hospital study conducted in Scotland even after 4 years exposure to potable water.[13]

Despite evidence of copper pipe corrosion in German hospitals, the exact cause has not been fully ascertained. Chamberlain et al.[24] have, however, shown that by using model polysaccharides, some of the copper corrosion events observed *in situ* could be studied and observed *in vitro*. They described similar results using exopolymeric substances produced from microorganisms isolated from perforated copper pipes. It has also been suggested that owing to the formation of an unequally distributed biofilm evident in potable water, local anodes form leading in some cases to the perforation of copper tubes. Therefore, the presence of copious amounts of exopolymeric substances and biofilm formation play a very crucial role in corrosion of copper in potable water environments.[24]

Cast iron pipes are also subjected to corrosion, often as a result of microbially induced corrosion owing to biofilm development. Corrosion of this nature often leads to coloured water and taste and odour problems which become an aesthetic problem to the consumer. However, on the public health side, corrosion of pipes is known to lead to a loss of chlorine residual, a depletion in oxygen levels, a reduction of sulphate to hydrogen sulfide, and a build up of iron precipitates in biofilms.[25,26] This, ultimately, generates an extensive ecosystem, protective haven, and niche for pathogens, but, in particular, faecal and total coliforms.

It is well documented that with the aging of a piping system comes accumulation of heavy scale corrosion products which leads to the restriction in the passage of water. This ultimately leads to pipeline breaks which causes bacterial colonisation and biofilm formation. Also, taste and odour problems may be generated and experienced by the consumer.

It has been extensively published that the major cited problems consumers experience include[25]

- Taste and odour—owing to algae,[27] bacteria (actinomycetes), iron/sulphate-reducing bacteria (*Desulfovibrio*),[28] and fungi.
- Red and cloudy water—owing to iron-oxidizing (*Gallionella* and *Leptothrix*) bacteria.[29]
- Black water—due to *Pseudomonas*, *Corynebacterium*, *Pedomicrobium*, and *Actinomycetes* as well as *Mycelia sterilia* (fungi imperfectants).[30]

10.3.2 The Mechanism of Microbially Induced Corrosion (MIC) in Potable Water

Corrosion is viewed as a series of electrochemical reactions at a metal surface when exposed to an aqueous phase-containing electrolyte. It was not until recently that biological processes were implicated in corrosion. Microbially induced corrosion (MIC) has generally been considered as occurring in anaerobic environments where sulphide-producing bacteria are active, but now aerobes have also been implicated in this corrosion process.[31]

Corrosion may occur on submerged metal surfaces owing to uneven colonisation of microorganisms.[32] As the microorganisms at the metal surface replicate forming microcolonies, an uneven distribution of developing microcolonies may occur resulting in areas of heavy and light colonisation. In areas of bulk fluids which are aerated, the oxygen consumption by the surface-attached bacteria can create an oxygen gradient near the metal surface. Therefore, the higher concentrations of oxygen occurs where the biofilm is in contact with the bulk aqueous phase and the lowest concentrations develop at the bottom of the biofilm which is in contact with the metal surface (Figure 10.1).[33] This type of oxygen concentration cell is likely to develop where an uncolonised area of the surface is in contact with the oxygenated bulk phase and meets an area covered by colonies of oxygen respiring bacteria. This would result in the area under the microcolony being anodic to the area exposed to the bulk aqueous phase (Figure 10.2).[33]

Microorganisms may also facilitate the formation of differential aeration cells on metal surfaces where an uneven distribution of corrosion products develop, induced by abiotic factors. A short period of time after initial colonisation of bacteria, other microcolonies of different species develop next to one another and merge to form a biofilm. As these biofilms contain a diverse range of microorganisms, a range of microbial activity will take place. The biofilm will restrict the diffusion of products

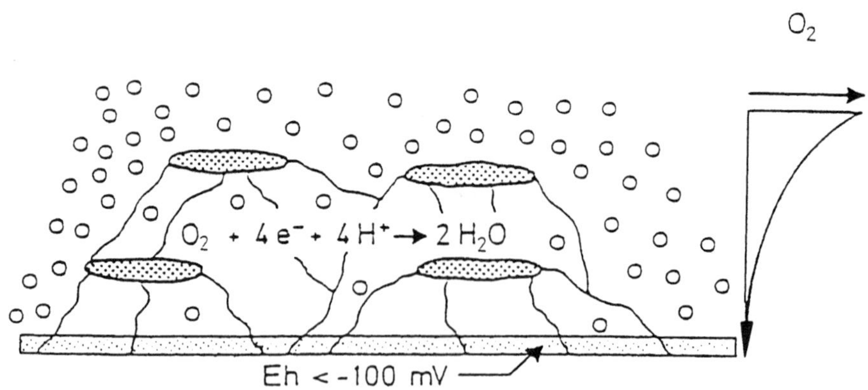

FIGURE 10.1 Oxygen concentration gradient in a biofilm caused by respiratory activity of microorganisms. From *Biofouling and Biocorrosion in Industrial Water Systems,* What is biocorrosion?, Flemming, H.C. and Geesey, G.G., Eds., 156, Copyright 1990, with permission from Springer-Verlag.

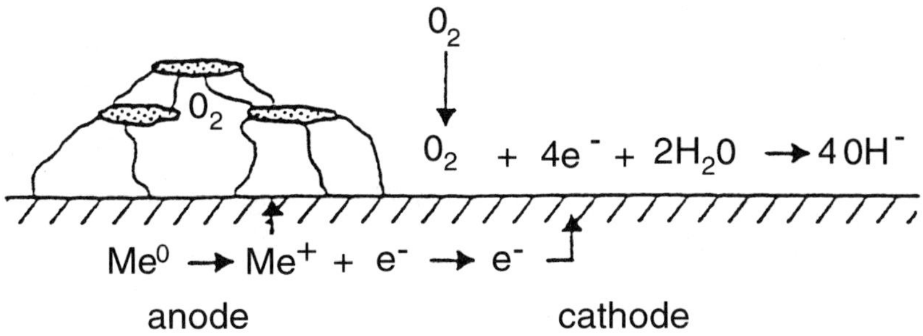

FIGURE 10.2 Differential aeration cell resulting from the heterogeneous distribution of biofilms. From *Biofouling and Biocorrosion in Industrial Water Systems,* What is biocorrosion?, Flemming, H.C. and Geesey, G.G., Eds., 156, Copyright 1990, with permission from Springer-Verlag.

of microbial metabolites excreted from cells in each microcolony. This will lead to very high concentrations of metabolic by-products at or near the underlying surface.[34] Where the surface is not compatible with the metabolites, corrosion may result.[35]

10.3.3 Characteristics of Materials in Potable Water

Before any material can be worthy of usage within potable water systems, it must show a high tolerance to the effect of microbial adhesion, biofouling, and biofilm development.[36,37] The physico-chemical and energy characteristics of many materials depend on its surface finish and surface roughness.[38] Surface roughness can be measured by the use of stylus type surface tracing which gives average peak to valley height (Ra). Today, aided by computers a three-dimensional surface profile can be obtained. Atomic force microscopy should make it possible to visualise microorganisms adhering to surfaces with varying degrees of surface roughness.[39-42]

When a material comes in contact with potable water, microbes adhere, grow, and produce extracellular material eventually forming a biofilm (Figure 10.3) of a complex microbial community. Eventually, the polymers produced by the microbes become immobilised within a three-dimensional matrix of microbes and organic ions. The chemical composition of this matrix is heterogeneous, often containing polysaccharides, proteins, and nucleic acids which help govern the properties and functions of the biofilm and, ultimately, its effects on corrosion.

Biofilms which form within potable water depend upon the hydrodynamic conditions of the system. Within static conditions, biofilms form as a result of bacterial sedimentation with nutrients regularly replenished within this type of system (dead legs). With a dynamic flowing system, biofilms are often thicker. However, the thickness of this biofilm will depend on the flow velocity of the liquid.[43-45] Therefore, as well as considering a material for its suitability for usage in potable water, the effects of biofilm development on its long term performance must be determined before it can be commissoned ultimately for use.

After several days of being immersed into potable water, it is possible to rank different materials according to the number of attached cells in potable water: cast

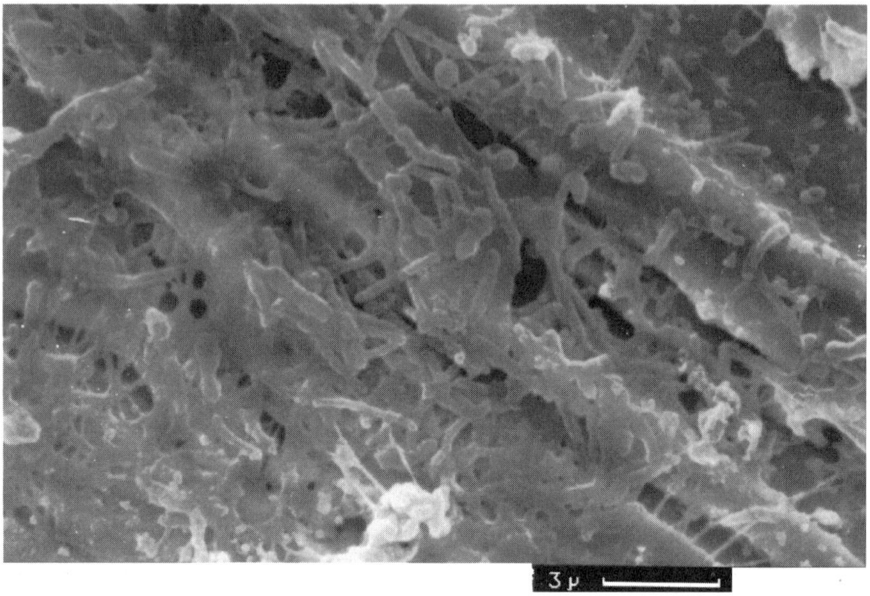

FIGURE 10.3 Biofilm in stainless steel potable water.

iron has a greater number of attached cells than does cement lined cast iron, which has a greater number of attached cells than does stainless steel.[46] Work completed recently has established that there are also significant differences between cast iron, medium density polyethylene (MDPE), and unplasticised polyvinyl chloride (uPVC) in drinking water.[47] Other research has found no significant differences in the accumulation of cells on copper, PVC, or iron surfaces,[48] but other work has contradicted this.[49]

Heat tint presence on the heat affected zone (HAZ) on some metal, but in particular stainless steel, has long been known to increase metal ion leaching and degradation of materials in high purity water.[50] Research work on MIC has also shown clearly that the removal of heat tint, by pickling or electropolishing, enhances resistance to both crevice corrosion and MIC. It now seems that the full inherent corrosion resistance of particular metals such as stainless steel is restored by preventing or removing heat tint.

10.4 NUISANCE ORGANISMS AND POTABLE WATER PROBLEMS

An important problem which is sometimes overlooked in piping materials is the fact that corrosion, as mentioned previously, can be induced by a number of microorganisms leading to the formation of MIC. The groups of microorganisms which bring about this effect can be classified as[51]

- Aerobic and facultative anaerobic heterotrophs.
- Autotrophic nitrifiers.
- Denitrifiers.
- Nitrogen fixers.

- Iron precipitating bacteria.
- Sulphate reducers.
- Sulphur oxidising.

It is acknowledged generally that in potable water there are a lot of microbes which constitute a nuisance rather than generating any public health concern. Therefore, problems associated with potable water must be found to be associated with biofilms rather than with free floating cells. A large variety of microorganisms have been found in biofilms attached to metal surfaces. The diversity of these microorganisms found within a biofilm which play a role in corrosion on these metal surfaces are shown in Table 10.1.[52]

The microbial composition of water samples and pipe corrosion deposits can be seen in Table 10.2.[51] The organisms in these biofilms range from viruses to complex multicellular organisms and can be divided into three main groups: algae, bacteria, and fungi.

TABLE 10.1
Microorganisms Associated with Corrosion

Microorganism	Growth Requirement	Corrosion and Related Problems
General aerobic microorganisms, e.g., *Aerobacter*, yeasts, moulds	Water carbon source, nitrogen and phosphorous, trace elements	Differential aeration cells
General anaerobic microorganisms	Water, carbon source, nitrogen and phosphorous, trace elements	Produces organic acids that preferentially chelate specific alloying elements
Iron-oxidizing bacteria	Water CO_2, oxygen, ferrous materials, nitrogen and phosphorus, trace elements	Oxidize ferrous (+2) to ferric (+3), cause blockage of pipes, create anaerobic conditions
Nitrite-oxidizing bacteria	Water, nitrate, carbon sources, phosphorus, trace elements, ammonia, and aeration	Consume nitrite corrosion inhibitors, differential aeration cell
Nitrate-reducing bacteria	Water, carbon source, nitrogen and phosphorus, sulphate, trace elements	Reduce nitrate, produce large quantities of organic acids
Pseudomonas sp.	Water, hydrocarbons, nitrogen and phosphorus, trace elements, manganese and iron, aeration	Skin infections, provide material for anaerobic organisms, differential aeration cell
Sulfur-oxidising bacteria	Water, sulphides or sulphur, CO_2, nitrogen and phosphorus, trace elements, oxygen	Produces up to 10% H_2SO_4, concrete attack
SRB	Water, carbon source, nitrogen and phosphorus, trace elements	Produce large quantities of sulfide, sever localized corrosion

Source: Adapted from Videla, H.E., *Manual of Biocorrosion*, ©1996, CRC Press, with permission. Reproduced with permission of Stein, A.A., MIC treatment and prevention, in *Practical Manual on Microbiologically Influenced Corrosion*, Kobrin, G., Ed., NACE International, Houston, TX, 1993, 101. NACE International is the copyright holder of this table.

TABLE 10.2
Microbial Composition of Water Samples and Pipe Corrosion Deposits

	Untreated Water (August (1989)		Untreated Water (March 1990)		Treated Water (March 1990)		Corrosion Tubercles	
Microorganisms	**20°C**	**8°C**	**20°C**	**8°C**	**20°C**	**8°C**	**20°C**	**8°C**
Aerobic SPC[a]	2.2 ¥ 10^{6}	1.5 ¥ 10^{5}	3.2 ¥ 10^{4}	9 ¥ 10^{3}	20	ND	2.9 ¥ 10^{7}	2.0 ¥ 10^{6}
Anaerobic SPC[a]	2.0 ¥ 10^{1}	ND	1.0 ¥ 10^{1}	ND	<1	ND	3.0 ¥ 10^{6}	ND
Total coliforms[b]	570	350	200	75	<1	ND	5.0 ¥ 10^{4}	ND
Fungal SPC[c]	4.8 ¥ 10^{4}	2.1 ¥ 10^{2}	3.5 ¥ 10^{3}	2.0¥10^{2}	3	ND	3.0 ¥ 10^{5}	7.8 ¥ 10^{2}
Iron reducers[d]	540	240	70	49	<0.3	<0.3	>24,000	430
Sulfate reducers[e]	280	130	<3	<3	<0.3	<0.3	280	110
Sulfite reducers[e]	120	93	4	4	<0.3	<0.3	460	210
Thiosulfate reducers[e]	540	170	240	79	9.3	1.5	920	540
Iron oxidizers[f]	54	24	7.9	7	<0.3	<0.3	75	64

[a] Cfu/ml (water) or cfu/g (corrosion tubercle), 7-day incubation at 20°C, 10-day incubation at 8°C.

[b] Cfu/ml (water) or cfu/g (corrosion tubercle), 48-day incubation at 35°C, 10-day incubation at 8°C.

[c] Cfu/ml (water) or cfu/g (corrosion tubercle), 7-day incubation at 20°C, 10-day incubation at 8°C.

[d] Organisms/ml (water) or cfu/g (corrosion tubercle), by 5-tube MPN, using dilutions of 1.0–0.0001 ml in B_{10} broth, 14-day incubation.

[e] Organisms/ml (water) or cfu/g (corrosion tubercle), by 5-tube MPN, using dilutions of 1.0–0.0001 ml in Butlin's broth and either 1% (v/v) Na_2SO_4, Na_2SO_3, or $Na_2S_2O_3$,14-day incubation.

[f] Organisms/ml (water) or cfu/g (corrosion tubercle), by 5-tube MPN, using dilutions of 1.0–0.0001 ml in modified Winogradsky's broth,14-day incubation.

Source: From Geldreich.[51]

10.4.1 Algae and Diatoms

Algae are mainly microscopic organisms, but some algae such as seaweeds are macroscopic. Algae are important fouling organisms in biofilms exposed to light because algae contain chloroplasts and are capable of photosynthesis. They are classified into seven divisions based on their morphology. The main groups of algae are green algae, red algae, brown algae, and diatoms. They play an important part in MIC owing to their ability to produce molecular oxygen, corrosive organic acids, slime, and nutrients for other microorganisms involved in MIC.[35] It is highly unlikely that these will be present in potable water in a viable state (owing to a lack of sunlight in the closed system).

Microalgae have been reported to affect electrochemical reactions directly. The diatom species, *Nitzchia*, has been shown to bring about a reduction in corrosion by causing cathodic polarisation.[35] Conversely, *Achnantles* was found to induce cathodic depolarisation. Of cyanobacterium, two species have been implicated in the corrosion of stainless steel.[32] As with macroscopic algae, diatoms are known not to proliferate in potable water owing to the restriction of a light source. However, diatoms are known to become entrapped in biofilms present in potable water. The number of different species present in these biofilm is quite large [Figures 10.4(a)–(e)]. Whether diatoms have any significance in potable water is presently an area which has not been investigated.

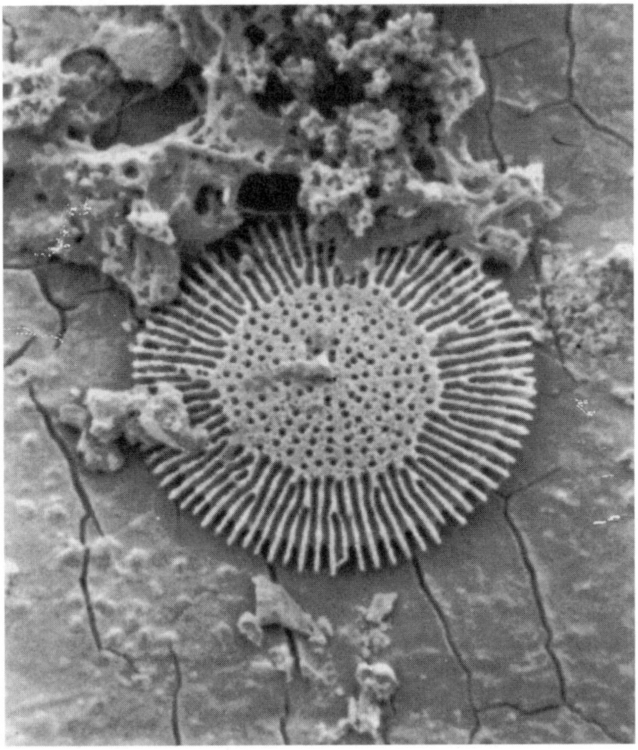

FIGURE 10.4a Diatom associated with surfaces in potable water.

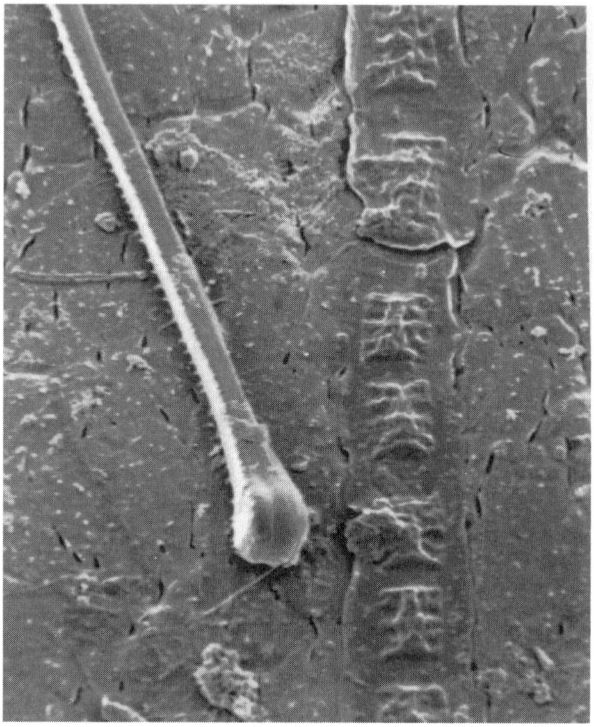

FIGURE 10.4b Diatom associated with surfaces in potable water.

10.4.2 Bacteria

10.4.2.1 Sulphate Reducing Bacteria

Bacteria are the primary organisms in most biofilms. Perhaps the most important biofilm bacteria associated with MIC are sulphate reducing bacteria (SRBs). The biochemistry and physiology of SRBs have been described by Postgate.[53] SRBs are anaerobic chemoheterotrophs that are physiologically and ecologically homogeneous. They derive their carbon and energy from organic nutrients and use sulphate as a terminal electron acceptor with the production of sulphide.[31] The production of hydrogen sulphide at the end of its metabolic pathway is generally found to be corrosive toward metal surfaces. SRBs are located deep in biofilms where anaerobic conditions prevail. They are known to utilise short chain organic acids released by other organisms present in a biofilm.[54] As these bacteria utilise the available organic nutrients, hydrogen sulphide is produced leading to the development of a foul odour.

SRBs constitute a very diverse range of organisms. The most commonly encountered organisms are *Desulfovibrio*, *Thiobacillus*, *Beggiatoa*, *Thiodendran*, and *Thiothrix*. However, sulphur oxidising bacteria, that is, *Thiobacillus* have been reported to have an important role in the association with microbiological fouling. Sulphur

FIGURE 10.4c Diatom associated with surfaces in potable water.

oxidising bacteria are often found in the association with SRBs, forming a consortium known as a sulphuretum, where oxidised and reduced sulphur environments will alternate.

10.4.2.2 Iron Bacteria

Iron precipitating bacteria are generally located in biofilms which produce tubercles in iron pipes leading to a reduction in water flow. Examples of these organisms include *Crenothrix polyspora* and *Sphaerotilus natans*. Others include *Gallionella*, *Hyphomicrobium*, and *Caulobacter*.

Iron oxidising bacteria obtain energy by oxidising ferrous ions (Fe^{2+}) to ferric ions (Fe^{3+}). They have been associated with the corrosion of water systems.[32] The main group of iron oxidising bacteria include the filamentous genus *Sphaerotilus*, its related forms, *Crenothrix* and *Leptothrix* species, and the unicellular stalked bacterium *Gallionella*. These bacteria are often found in association with sulphate reducing and oxidising bacteria.[55] *Pedomicrobium manganicum* has also been shown to be able to oxidise ferrous iron on stainless steel.[32] It also has been known to oxidise manganese in potable water systems and bind MnO_2 in its extracellular polysaccharides.[56]

FIGURE 10.4d Diatom associated with surfaces in potable water.

Iron reducing bacteria include *Pseudomonas* sp. which have been implicated in the reduction of Ferric (Fe^{3+}) to ferrous iron (Fe^{2+}). Bacteria capable of reducing Fe^{3+} anoxically include *Bacillus*, *Pseudomonas*, *Micrococcus*, *Corynebacterium*, *Alcaligenes*, and *Vibrio*, to name but a few.[57] Particularly, *Pseudomonas* and *Bacillus* sp. are the most common iron oxidising bacteria found in water distribution systems.[37]

10.4.2.3 Nitrogen Bacteria

Nitrogen utilising bacteria are involved in the recycling of nitrogen compounds. Aerobic species, such as *Nitrosomonas*, will oxidise ammonia to nitrite which in turn is oxidised to nitrate by *Nitrobacter* spp. These organisms rarely exist in isolation and are usually part of a complex consortial biofilm.

10.4.2.4 Manganese Utilising Bacteria

Manganese utilising bacteria generally lead to the creation of black water. Organisms which have been located in a biofilm that ultilises manganese (greater than 0.01 to 0.5 mg per litre) include *Pseudomonas*, *Arthrobacter*, *Hyphomicrobium*, and *Sphaerotilus discophorus*.[58]

FIGURE 10.4e Diatom associated with surfaces in potable water.

10.4.2.5 Fungi

Microfungi grow saprophytically on nonliving organic material deriving energy from fermentation or oxidation. Organic acids, produced as an end product in fungal biochemical reactions, are generally corrosive to metal surfaces particularly when associated with biofilms.

10.5 CHARACTERISTICS OF MATERIALS WHICH HAVE EFFECTS ON BIOFILM FORMATION

The welded joints of metal pipes found in potable water have enormous potential for bacterial adhesion and, ultimately, biofouling, owing the large surface area available for attachment (Figure 10.5).

Recent research into the factors influencing microbiological corrosion of stainless steel indicates that welding reduces the resistance of stainless steel against MIC. Punter[50] found no cases of stainless steel corrosion on stainless steel samples without a weld. The usage and performance of stainless steel in potable water may be largely determined by the way in which the pipe surfaces are cleaned and disinfected. Optimisation of the cleaning and disinfecting procedures appears to be a way to achieve control over MIC of stainless steel.[38]

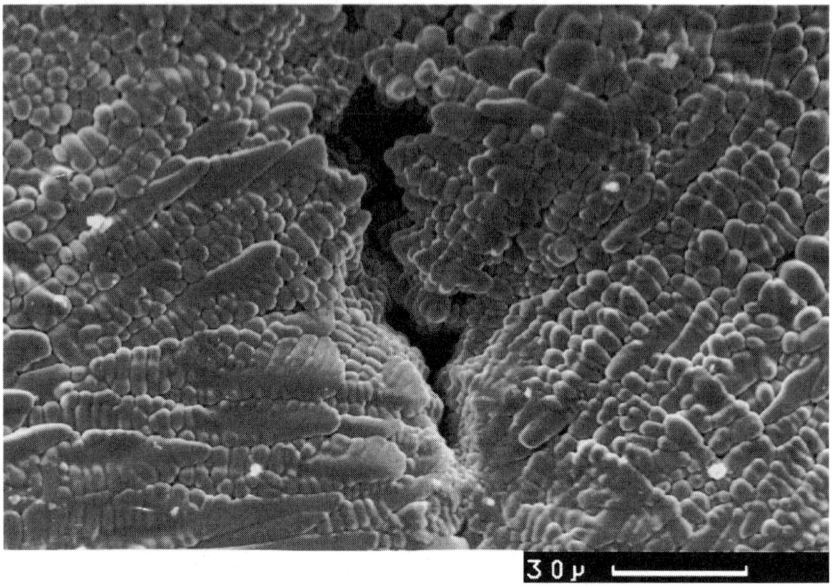

FIGURE 10.5 Weld showing the presence of large pores which aid in bacterial attachment.

When materials are used in potable water environments for the first time, MIC must be viewed as a possibility. However, provided that specific guidelines and certain measures are undertaken, metallic materials used in potable water have been known to have very long service lives.

10.6 THE USE OF STAINLESS STEEL IN POTABLE WATER

Domestic potable water installations are part of the distribution system which ends at the consumer's tap. Many decades ago, concerns were mainly owing to the technical quality of the materials and the plumbing system with regard to failures by water hydraulics, whereas water quality supplied to the consumer was not a major concern. However, priorities have changed in Europe within the last decade owing to the EC directive on Potable Water and the National Potable Water Regulations. Although stainless steel is not a standard material for distribution systems, it has acted as a materials solution in several problem areas. These have included pipe attachments to bridge spans, trenchless piping, New York City risers, and laterals. Other examples are those in Japan, Sweden, and Italy where PVC and ductile mains water pipes are being replaced by grade 316 stainless steel following 10 years of testing.

During the 1960s, new homes in numerous counties of England were plumbed with type 302 (UNS S30200) light gauge stainless steel tubing in lieu of copper (which was more expensive at the time). These installations were made with the approval of the British Water Works Association. The principal producer of stainless steel pipe in England estimated that approximately 16.4 million feet of pipe used to transport potable water was placed in service during the 1966 to 1982 period. Tuthill,[59] through extensive surveys, found that 5 counties that had stainless steel pipe in service for 20 years or more had no problems with stainless steel pipe failure and corrosion. This study established that stainless steel could be used as a long term viable piping system for transporting potable water. However, owing to high stainless steel prices at the time, it was never fully commisioned as a cost-effective alternative to copper for transporting potable water.

Today, areas in Tokyo, New York, and the Middle East have used stainless steel for certain areas of their potable water distribution systems, reestablishing evidence on the long term performance and sustainability of stainless steel. In Tokyo, after a period of 10 years of evaluating candidate materials, the Water Works Bureau of Tokyo (WBT) in 1980 began replacing leaking connectors, joining the street sub-mains to the dwellings outside meters, with 1 inch (internal diameter) type 316 piping.[60] After 11 years of use, there have been no reported internal or external corrosion of any of the stainless steel connectors. System leakage had also been dramatically reduced.[61]

In the U.S., the Bureau of Water Supply in New York, after extensively evaluating materials for use in the Rondout Reservoir, selected 304 stainless steel pipes for valves in piping systems.[61] After a period of 6 years, only 1 instance of corrosion had been reported with these valves, which was owed to water being left stagnant in the pipeline for several months.

In the Middle East, purified water from desalination plants is blended with local groundwater for potable use. The piping used to transport this water is welded, large diameter stainless steel, typically 24 inch, and usually grade 316L. Some leakage problems were recorded in three of these plants, owing to an uncontrolled usage of hypochlorite which had been left standing in the piping system. These failures, as in New York, were attributed to overchlorination.[59]

Stainless steel piping has been used instead of ductile iron in more than 30 potable water treatment plants in the U.S., largely because of the financial savings.[61] Savings from the use of stainless steel over ductile iron were estimated at $50,000 for the Tauton potable water plant in Massachusetts.[62]

Therefore, the performance of stainless steel in potable water has led to an increase in consumer confidence for usage. With a good understanding of fabrication and post fabrication of stainless steel, combined with a knowledge of the effects of chloride levels, oxidants, and conditions which may lead to both crevice and MIC, it is feasible that stainless steel has a service life of 100 years.[63]

10.6.1 Composition and Physical Characteristics of Stainless Steel

There are a number of grades of stainless steel with different compositions and properties and, consequently, different uses. Stainless steels are alloy steels which in addition to iron, contain chromium, nickel, molybdenum (grade 316), and small amounts of other elements. This can be seen in Table 10.3. Types 304 and 304L are the most widely used basic grades of chromium-nickel stainless steels which are

TABLE 10.3
Chemical Composition (%/wt) of Stainless Steel 304 and 316 Pipe Sections (Avesta Sheffield Ltd.)

Elements	Pipe 304	Pipe 316
C	0.054	0.022
Si	0.420	0.410
Mn	1.860	1.500
P	0.023	0.024
S	0.0120	0.002
Cr	18.20	16.730
Mo	0.170	2.040
Ni	10.190	10.980
Nb	0.010	0.010
Co	0.110	0.160
Cu	0.150	0.190
Sn	0.010	0.012
W	0.040	0.130
N	0.079	0.049
V	0.050	0.050

used for potable water applications. Types 316 and 316L are the more corrosion-resistant grades which, in addition to chromium and nickel, contain molybdenum and are shown to be a better performer in potable waters than the 304 grades.[64]

Stainless steels have an oxide film on the surface, composed of oxy-hydroxide of chromium and iron,[65] and are referred to as a passive film. This makes the steel resistant to corrosion in aggressive, acid, or neutral-chlorinated solutions. The composition of the passive layer depends on the metal substratum, the surface finishing treatment (annealing or pickling), and the medium in which it is immersed.[66] Whilst the passive film present is a tough durable oxide layer, it occasionally may contain defects. It is at such defects that stainless steel may corrode when environmental conditions become aggressive enough to take advantage of any weakness in the film. Stainless steel grades 304 and 316, with a 2B finish, have been found to exhibit a surface oxide layer containing mainly chromium, oxygen, iron, and carbon. Energy dispersive X-ray analysis and electron spectroscopy for chemical analysis of the surfaces of stainless steel grades 304 and 316 have confirmed the presence of chromium, manganese, iron, and nickel in the oxide layer with molybdenum identified only on the surface of stainless steel grade 316.[64]

In the continued effort to enhance corrosion resistance, elements such as nickel, molybdenum, and titanium have been added to stainless steel. The result is alloys with strength and corrosion resistance best demonstrated by the 300 series.[67] Molybdenum greatly enhances resistance to localised corrosion which gives grade 316 steel an advantage over grade 304. The low levels of carbon present in grades 304

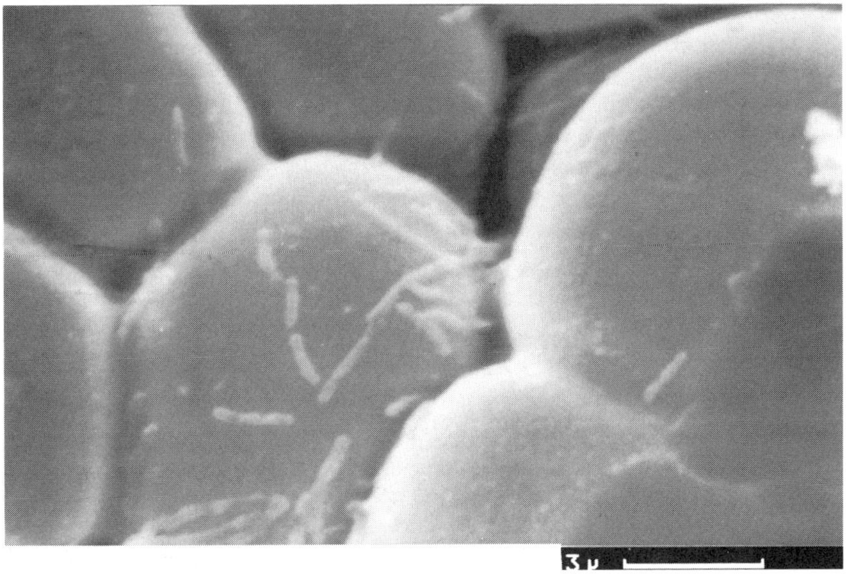

FIGURE 10.6 Small colonies of bacteria observed on stainless steel after exposure to potable water.

and 316 stainless steel should not give rise to significant problems associated with welding of these alloys. However, special low carbon (less than 0.03%) grades, 304L and 316L, are available for certain welding applications and provide greater resistance to corrosion in potable water.

On stainless steel, two surface finishes, available commercially, are generally used in the transport of potable water. These include the 2B pickled finish and the 2R bright annealed finish. However, a 2D surface finish may also be used (Table 10.4). Stainless steel is obtained industrially by direct continous casting of slabs, which are hot-rolled. Following this process, the stainless steel sheets are subsequently annealed and then cold-rolled into a tube. A final annealing follows and, in some cases, the sheet is pickled if the annealing has been done in an oxidising atmosphere (the surface quality is then designated 2D finish). To avoid the final pickling process, the steel is annealed in a protective atmosphere, giving a bright

TABLE 10.4
Surface Finishes Available on Stainless Steel (from BS 1449: Part 2)

Finishes	Description	Comments
2D	Cold rolled, softened, and descaled	A matt finish for general applications
2B	Cold rolled, softened, descaled, and lightly rolled on polished rolls	A smooth finish for general applications, brighter than 2D finish
2A	Bright annealed	A cold-rolled reflective finish retained through annealing

finish designated 2R in France and BA in the U.S. With the 2D finish, a skin-pass operation is carried out to enhance the final brightness, the surface quality so obtained being designated 2B or pickling finish. To achieve finish 4, 2D, or 2R pipe surfaces sheets are polished with fine-grained polishing belts. For stainless steel, both basic surface composition and roughness differ according to the type of finish required with roughness increasing with the thickness of the metal sheet and tube.[68] The surface finish determines the energy characteristics of stainless steel which will have very important implications on the adhesion rate of potable water bacteria and pathogens. However, the role of surface finish with respect to bacterial adhesion has not yet been determined clearly, but it may be assumed that stainless steel with the smoothest finish would have the lower initial bacterial adhesion rate in potable water.[14]

10.6.2 A Comparison of Biofilm Development on Stainless Steel Grades 304 and 316 and Other Materials in Potable Water

Recent work on the use of stainless steel in potable water supplies has established some very interesting results suggesting differences between stainless steel grades 304 and 316 commissioned for use in potable water.[13,14,42,64] It is well known that metallic surfaces play an important role in the early stages of biofilm development, influencing both the rate of microbial adhesion and the distribution of these cells.[69] An area of particular importance is surface roughness where increasing surface roughness generally increases microbial colonisation.[70–72] It has also been found that porous welds, an observed characteristic of stainless steel 304 pipe weld, may provide increased sites for colonisation and ultimately biofilm formation.[13,73,64]

Scanning electron micrographs taken of the surface of stainless steel following exposure to potable water have confirmed that microorganisms accumulate at discontinuities on the submerged stainless steel surfaces.[64,72,74] The effect of surface roughness on surface conditioning (organic) and on subsequent adhesion and biofilm formation on stainless steel in potable water is of interest. Because all surfaces immersed in potable water environments are rapidly coated with a biological layer (conditioning film), it is presumed that both smooth and roughened surfaces will be coated. However, the hydrophobicity of the surface, the nature of the environment, and the degree and type of surface rugosity may well determine the nature and amount of conditioning film. The surface might be coated with a thin conditioning film which mirrors surface irregularities or the coating might be uneven, related to flow in a moving system. These features may well influence the speed of microbial deposition and the type of organisms colonising surfaces and, therefore, may have an effect on the corrosion rates of these two stainless steel grades particularly in potable water.

Previous research on biofilm formation on stainless steel in potable water has shown that matt stainless steel (of unknown grade, which was characterised as rough) accumulated 1.44 times more microbes than electropolished stainless steel.[44,69] This would suggest that stainless steel with a smooth surface finish decreases the adhesion of microorganisms.

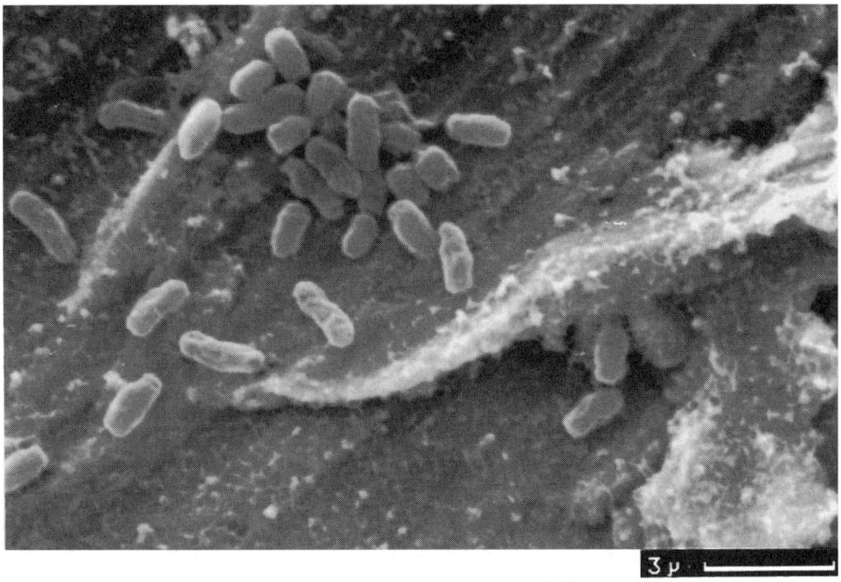

FIGURE 10.7 Evidence of small and single colonies in potable water.

In all potable water-based studies colonisation of bacteria on the stainless steel surfaces observed using epifluorescence and scanning electron microscopy has been found to be patchy, consisting of small colonies (Figure 10.6) and single cells (Figure 10.7). There seems to be no uniform microbial colonisation of either stainless steel grades in potable water. Scanning electron microscopy has revealed several types of sessile microorganisms including rod-shaped bacteria, coccoid bacteria, spiral bacteria, filamentous bacteria, curved rods, yeast cells, many different species of diatoms, and fungal hyphae attached to stainless steel in potable water. These microbial colonies are generally found to be present in crevices and fissures on the stainless steel surfaces, particularly at high velocities.

From studies looking at stainless steel in potable water, it is generally found that grade 304 is colonised by bacteria at a higher rate than grade 316. Studies have also found that molybdenum, present in the passive layer of grade 316, has an effect on bacterial growth in potable water by reducing the viability and biofilm development of heterotrophic and sulphate reducing bacteria.[75-77]

10.7 EFFECTS OF DISINFECTANTS ON STAINLESS STEEL AND OTHER MATERIALS IN POTABLE WATER

Chlorination of potable water is standard practice, providing excellent protection against bacterial strains that cause either public health problems or microbiologically induced corrosion. Also, the material chosen to distribute potable water must not have effects on demand for chlorine because this will have effects on the disinfection process and, ultimately, reduce chlorine availability to treat water. Within iron pipes, chlorine doses as high as 4 mg per litre have been shown to have little impact on biofilm reduction because iron corrosion products interfere with free chlorine disinfection. It is often found that the chlorine demand is 10 times higher in iron pipes than other pipes of other composition.[78,79]

Other considerations that need to be taken into account for the use of materials used in potable water is the effect chlorine may have on increasing the potential to cause corrosion on certain metals. For example, it is often documented that chlorine has to be administered with care when using stainless steel so as not to leave too high a chorine residual known to have effects on its corrosion potential. There are situations in which corrosion (MIC) appears to have occurred on stainless steel, owing to high levels of residual chlorine. These have been identified in water treatment plants on the welds on 304L stainless steel piping. This seems to be owing to hydro-test water being left standing in the stainless steel piping for a month or more, allowing bacteria to induce microbiologically induced corrosion.[59] Type 304 stainless steel is generally acceptable for chloride levels up to 200 mg per litre chloride and 316 for levels up to 1000 mg per litre.

Sedimentation water found within crevices (such as those that originate from incomplete fusion of the circumferential weld) of stainless steel under some conditions can lead to localised pitting-type attacks on stainless steel if chlorides are present in sufficient concentrations. Crevice corrosion of type 304 stainless steel is rare at below 1000 mg per litre chlorides.[80] However, crevice corrosion can occur in water of lower chloride content, particularly if a mechanism of chlorine concentration is present. With respect to copper, the highest corrosion probability is observed for intensively branched, horizontally installed pipework with prolonged periods of stagnation.[20]

Research has shown that low carbon grades of stainless steel—types 304L and 316L avoid corrosion in the heat affected zone (HAZ) of welds at the 1 to 2 mg per litre residual chlorine concentrations normally encountered in potable water. Type 316/316L is resistant up to 5 mg per litre residual chlorine. Studies indicate that type 316L would be a safer choice than type 304L when higher residual chlorine is encountered.[59]

When chlorine is added to manganese-bearing waters with oxidising bacteria such as *Gallionella*, a self-sustaining corrosion reaction is initiated, resulting in severe pitting of stainless steel.[81] Although corrosion of this type has not been reported in potable water plants, it is possible and should be considered when raw waters contain appreciable manganese.

10.8 CONCLUSION

Pipe materials are the most extensive part of any potable water system. The characteristics of a piping material are fundamental to the development of a biofilm. Other factors which need to be taken into account when evaluating materials for the transport of potable water are the length of the pipe needed, the predominant pipe material, the age of the pipe, corrosion, and zones of static water. Whilst cast iron has been in use for over 280 years, the choice of materials used to transport potable water are immense. The choice of material used to transport potable water lies with respect to its application. For example, copper would never be used in a water distribution system. However that was often said of stainless steel, but now it is being used as a piping material in water distribution particularly in Turin, Italy. However as a general understanding, the choice of material for long runs of distribution pipe include ductile iron, cast iron lined with cement, steel, reinforced concrete, and PVC. Within service lines and building pipe networks, the most predominately used materials are plastics and copper. However, stainless steel is beginning to replace traditional materials such as these.

Overall, stainless steel is a very worthy candidate for the replacement of copper as a domestic and industrial potable water supply pipe system, particularly in areas where corrosion is proliferate. They have extremely low corrosion rates in potable water and are 100% recyclable. Stainless steel can handle turbulent high velocity waters and provide a long service life providing proper operational repair is applied. However, research on the usage and overall performance of stainless steel in potable systems is very limited, suggesting this is a very important area for future investigations.[13] More research is necessary on metal ion leaching from stainless steel into potable waters, particularly owing to changes in the EC standards for metal ions within potable water, and the effects of residual chlorine on the corrosion potential of stainless steel. Also, all aspects of biofilm development, particularly consortial development, paying particular attention to pathogenic and coliform inhibition, is a necessity if this metal's overall performance is going to be fully accepted for use in potable water.

Stainless steel is a very cost effective material of construction when compared with the more traditional construction materials which are used in potable water.

Guidelines are available for the installation and usage of stainless steel within institutional buildings' plumbing systems, particularly hospitals.[9] This suggests that the use of stainless steel within municipal buildings' plumbing systems is becoming more widespread and dependable, confirming stainless steel as a reliable alternative to the normal traditional materials such as copper.

10.9 REFERENCES

1. Geldreich, E. E., 1996, Microbial quality of water supply in distribution systems, Lewis Publishers, New York.
2. Houston, A., 1916, The growth of microbes on leather washers, 12th Research Report, Metropolitan Water Board.
3. Committee Report, 1979, The use of plastics in distribution systems, *J. Am. Water Works Assoc.*, 71, 373.
4. Wagner, I., 1992, Internal corrosion in domestic drinking-water installations, *J. Water SRT-Aqua*, 41, 219.
5. Institute of Plumbing, 1988, *Plumbing Engineering Services Design Guide,* Institute of Plumbing, Hornchurch.
6. National Association of Corrosion Engineers, 1980, Prevention and control of water caused problems in building potable water systems, Publication TPCX 7, National Association of Corrosion Engineers, Houston, TX.
7. **Anon.,** 1987, Service water heating, in *HVAC Handbook,* American Society of Heating, Refrigeration, and Air Conditioning Engineers, Atlanta, GA, chap. 54, 6.

8. Davis, A. R., 1951, The distribution system, in *Manual for Water Works Operators,* Texas State Dept. Health and Texas Water and Sanitation Research Foundation, Austin, TX, 342.
9. **Anon.,** 1994, Domestic hot and cold water systems for Scottish health care premises, Scottish Hospital Technical Note 2, Her Majesty's Stationary Office, London.
10. World Health Organisation, 1990, European conference on environment and health: the European charter and commentary, *Proc. First European Conf. Environment and Health,* December 7–8, 1989, World Health Organisation, Frankfurt, 1990.
11. Burman, N. P. and Colbourne, J. S., 1977, Techniques for the assessment of growth of microorganisms on plumbing materials used in contact with potable water supplies, *J. Appl. Bacteriol.,* 43, 137.
12. Schwenk, W., 1991, Nickel migration from Cr-Ni stainless steel exposed to potable water, *Br. Corros. J.,* 26, 245.
13. Percival, S. L., Beech, I. W., Edyvean, R. G. J., Knapp, J. S., and Wales, D. S., 1997, Biofilm development on 304 and 316 stainless steels in a potable water system, *Inst. Water Environ. Manage.,* 11, 289.
14. Percival, S. L., Knapp, J. S., Wales, D. S., and Edyvean, R., 1998, Biofilm development on stainless steel in mains water, *Water Res.,* 32, 243.
15. Lewus, M. O., Hambleton, R., Dulieu, D., and Wilby, R. A., 1999, Behaviour of ferritic, austenitic and duplex stainless steels with different surface finishes in tests for metal release into potable waters based upon the procedure BS7766:1994, in *Stainless Steel '99 Science and Market, Proc.,* Vol. 1, *Marketing and Application,* 3rd European Congress, Chia Laguna Sardinia, Italy, 387.
16. Nuttall, J. C., 1993, in *Corrosion and Related Aspects of Materials for Potable Water Supplies,* McIntyre, P. and Mercer, A. D., Eds., The Institute of Materials, London, 65.
17. Voutilainen, P., 1990, Overview of developments in the copper industry, *Copper '90: Fining, Fabrication, Markets,* The Institute Of Metals, Bourne Press, Bournemouth, Dorset.
18. McEvoy, J. and Colbourne, J. S., 1988, Glasgow Hospital Survey Pitting Corrosion of Copper Tube, Report to International Copper Research Association, New York.
19. Keevil, C. W., Walker, J. T., McEvoy, J., and Colbourne, J. S., 1989, Detection of biofilms associated with pitting corrosion of copper pipework in Scottish hospitals, in *Biocorrosion,* Gaylarde, L. C. and Morton, L. H. E., Eds., Biodeterioration Society, Kew, England, 99.
20. Fischer, W., Paradies, H. H., Wagner, D., and Haenssel, I., 1992, Copper deterioration in a water distribution system of a county hospital of Germany caused by microbially induced corrosion. Part 1: description of the problem, *Werkst. Korros.,* 43, 56.
21. Shalaby, H. M., Al-Kharafi, F. M., and Gouda, V. K., 1989, A morphological study of pitting corrosion of copper in soft tap water, *Corrosion,* 45, 536.
22. Walker, J. T., Dowsett, A. B., Dennis, P. J. L., and Keevil, C. W., 1991, Continous culture studies of biofilm associated with copper corrosion, *Int. Biodeterior.,* 27, 121.
23. Wagner, D., Fischer, W., and Paradies, H. H., 1992, First results of a field experiment in a county hospital in Germany concerning the copper deterioration by microbially induced corrosion, in *Proc. 2nd Int. EFC Workshop on Microbial Corrosion,* 243.
24. Chamberlain, A. H. L., Fischer, W. R., Hinze, U., Paradies, H. H., Sequeira, C. A. C., Siedlarek, H., Thies, M., Wagner, D., and Wardell, J. N., 1995, An interdisciplinary approach for microbially influenced corrosion of copper, in *Proc. 3rd Int. EFC Workshop,* Tiller, A. K. and Sequeira, C. A. C., Eds., European Federation, The Institute of Materials, 1.

25. O'Connor, J. T., Hash, L., and Edwards, A. B., 1975, Deterioration of water quality in distribution systems, *J. Am. Water Works Assoc.*, 67, 113.
26. Banerji, S. K., Knocke, W. R., Lee, S. H., and O'Connor, J. T., 1978, Biologically mediated water quality changes in water distribution systems, EPA Project Report, Cincinnati, OH.
27. Osborn, E. T. and Higginson, E. C., 1954, Biological corrosion of concrete, Joint Report Field Crops Research Branch, Agriculture Research Series, U.S. Dept. of Agriculture and Bureau of Reclamation, U.S. Dept. of the Interior.
28. Mackenthun, K. M., 1969, The practice of water pollution biology, U.S. Department of the Interior, U.S. Government Printing Office, Washington, D.C.
29. Lueschow, L. A. and Mackenthun, K. M., 1962, Detection and enumeration of iron bacteria in municipal water supplies, *J. Am. Water Works Assoc.*, 54, 751.
30. Schweissfurth, R., 1978, Manganese and iron oxidizing microorganisms, *Landwirsch. Forsch.*, 31, 127.
31. Hamilton, W. A., 1985, Sulphate-reducing bacteria and anaerobic corrosion, *Ann. Rev. Microbiol.*, 39, 195.
32. Iverson, W. P., 1987, Microbial corrosion of metals, *Adv. Appl. Microbiol.*, 32, 1.
33. Geesey, G. G., 1990, What is biocorrosion?, in *Biofouling and Biocorrosion in Industrial Water Systems*, Flemming, H. C. and Geesey, G. G., Eds., Springer-Verlag, New York, 155.
34. Walker, 1990.
35. Ford, T. and Mitchell, R., 1990, Ecology of microbial corrosion, *Adv. Microb. Ecol.*, 11, 231.
36. Donlan, R. M. and Pipes, W. O., 1986, Pipewall biofilm in drinking water mains, 14th Annual AWWA Water Quality Technology Conference, Portland, OR.
37. Emde, K. M., Smith, D. W., and Facey, R., 1992, Initial investigation of microbially influenced corrosion in a low temperature water distribution system, *Water Res.*, 26, 169.
38. Haudrechy, P., Petermann-Boulange, L., Fontaine-Bellon, M. N., and Baroux, B., 1993, Wettability and bacterial adhesion on stainless steel: the respective effect of the surface condition and the cleaning processes, Innovation of Stainless Steel, Florence, Italy, October 11–14, 1993, 2169.
39. Steele, A., Goddard, D. T., and Beech, I. B., 1994, An atomic force microscopy study of the biodeterioration of stainless steel in the presence of bacterial biofilms, *Int. Biodeterior. Biodegr.*, 34(1), 35.
40. Beckman, M., Kolb, H. A., and Lang, F., 1995, Atomic force microscopy of biological cell membranes: from cells to molecules, *Eur. Microsc. Anal.*, 1, 5.
41. Surman, S. B., Walker, J. T., Goddard, D. T., Morton, L. H. G., Keevil, C. W., Weaver, W., Skinner, A., and Kurtz, J., 1996, Comparison of micrscope techniques for the examination of biofilms, *J. Microbiol. Meth.*, 25, 57.
42. Percival, S. L, Knapp, J. S., Wales, D. S., and Edyvean, R. G. J., 1998, Biofilms, mains water and stainless steel, *Water Res.*, 32(7), 2187.
43. Pedersen, K., 1982, Method for studying microbial biofilms in flowing water systems, *Appl. Environ. Microbiol.*, 43, 6.
44. Pedersen, K., 1990, Biofilm development on stainless steel and PVC surfaces in drinking water, *Water Res.*, 24, 239.
45. Mittleman, M. W., Nivens, D. E., Low, C., and White, D. C., 1990, Differential adhesion, activity and carbohydrate: protein ratios of *Pseudomonas atlantica* monocultures attaching to stainless steel in a linear shear gradient, *Microbiol. Ecol.*, 19, 269.

46. Holden, B., Greetham, M., Croll, B. T., and Scutt, J., 1995, The effect of changing inter process and final disinfection reagents on corrosion and biofilm growth in distribution pipes, *Water Sci. Tech.,* 32, 213.
47. Kerr, C. J., Osborn, K. S., Robson, G. D., and Handley, P. S., 1997, The effect of substratum on biofilm formation in drinking water systems, in *Biofilms: Community Interactions and Control,* Wimpenny, J., Handley, P., Gilbert, P., Lappin-Scott, H., and Jones, M., Eds., Third Meeting of the British Biofilm Club, Gregynog Hall, Powys, 167.
48. O'Conner, J. T. and Banerji, S. K., 1984, Biologically mediated corrosion and water quality deterioration in distribution systems, EPA Report 600/2/84/056, Cincinnati, OH, 442.
49. Walker, J. T., Sonesson, A., Keevil, C. W., and White, D. C., 1993, Detection of *Legionella pneumophila* in biofilms containing a complex microbial consortium by gas chromatography–mass spectrometry analysis of genus-specific hydroxy fatty acids, *FEMS Microbiol. Lett.,* 113, 139.
50. Punter, A., 1994, Influence of weld discoloration on the susceptibility of stainless steel weldments to microbiologically influenced corrosion, *Stainless Steel Eur.,* October, 38.
51. Geldreich, E. E., 1996, *Microbial Quality of Water Supply in Distribution Systems,* CRC Press, Boca Raton, FL.
52. Videla, H. E., 1996, *Manual of Biocorrosion,* CRC Press, Boca Raton, FL.
53. Postgate, J. R., 1984, *The Sulphate-Reducing Bacteria,* 2nd ed., Cambridge University Press, Cambridge.
54. Hamilton, W., 1990, Sulphate-reducing bacteria and their role in microbially influenced corrosion, in *Microbially Influenced Corrosion and Biodegradation,* Dowling, J. E., Mittleman, M. W., and Danko, J. C., Eds. University of Tenessee, Knoxville.
55. Tatnall, R. E., 1981, Case histories: bacteria induced corrosion, *Mater. Perform.,* 20, 32.
56. Sly, L. I., Arunpairojana, V., and Dixon, D. R., 1990, Binding of colloidal MnO_2 by extracellular polysaccharides of *Pedomicrobium manganicum, Appl. Environ. Microbiol.,* 56, 2791.
57. Ghiorse, W. C., 1988, Microbial reduction of manganese and iron, in *Biology of Anaerobic Microorganisms,* Zehnder, A. J. B., Ed., Wiley InterScience, New York, 305.
58. Committee on the Challenges of Modern Society (NATO/CCMS), 1987, Drinking water microbiology, *J. Environ. Pathol. Toxicol. Oncol.,* 7, 1.
59. Tuthill, A. H., 1994, Stainless-steel piping, *AWWA,* July, 67.
60. Sekine, Y., 1990, Water supply in fast-growing cities—Tokyo, Japan, *Aquas.,* 39, 86.
61. Tuthill, A. H. and Avery, A. E., 1994, Survey of stainless steel performance in low chloride waters, *Public Works,* 125, 49.
62. Tuthill, A. H., 1993, Save $50,000 using stainless steel instead of ductile iron, *Nickel,* 5, 4.
63. Powell, C. A. and Lamb, S., 1999, Stainless steels and their use in water treatment and distribution, in *Stainless Steel '99 Science and Market, Proc.,* Vol. 1, *Marketing and Application,* 3rd European Congress, Chia Laguna Sardinia, Italy, 373.
64. Percival, S. L., Knapp, J. S., Wales, D. S., and Edyvean, R., 1998, Physical effects on bacterial fouling of types 304 and 316 stainless steels, *Br. Corros. J.,* 33, 121.
65. Okamoto, S., 1973, Passive film of 18-8 stainless steel structure and its function, *Corros. Sci.,* 13, 471.

66. Barouxeranger, B., Beranger, G., and Lemaitre, C., 1993, Passivity and passivity breakdown on stainless steels, in *Stainless Steels*, Lacombe, P., Baroux, B., Beranger, G., Eds., Les Editions de Physique, Les Ulis, France, 163.
67. Tarara, J., 1988, Stainless steel gaining ground on corrosion, *Water/Eng. Manage.*, January, 33.
68. Bouhier, M. P., 1976, Etais de livraison des produits plats inoxydables comme critere de choix en fonconostoc de l'exigenece de l'etat de surface final, *Aciers Speciaux*, 33, 21.
69. Jain, D. K., 1995, Microbial colonization of the surface of stainless steel in a deionized water system, *Water Res.*, 29, 1869.
70. Geesey, G. G. and Costerton, J. W., 1979, Microbiology of a northern river: bacterial distribution and relationship to suspended sediment and organic carbon, *Can. J. Microbiol.*, 25, 1058.
71. Characklis, W. G., 1984, Biofilm development: a process analysis, in *Microbial Adhesion and Aggregation,* Marshall, K. C., Ed., Springer, New York, 137.
72. Geesey, G. G., Gillis, R. J., Avci, R., Daly, D., Hamilton, W. A., Shope, P., and Harkin, G., 1996, The influence of surface features on bacterial colonisation and subsequent substratum chemical changes of 316L stainless steel, *Corros. Sci.*, 38, 73.
73. Little, B. J., Wagner, P., and Jacobus, J., 1988, The impact of sulfate reducing bacteria on welded copper-nickel seawater piping systems, *Mater. Perform.*, 27, 56.
74. Walsh, D., Pope, D., Danford, M., and Huff, T., 1993, The effect of microstructure on microbiologically influenced corrosion, *J. Min. Met. Mat. Soc.*, 45, 22.
75. Beech, I. B. and Cheung, C. W. S., 1995, Interactions of exopolymers produced by sulphate-reducing bacteria with metal ions, *Int. Biodeterior. Biodegrad.*, 35, 59.
76. Chen, G., Ford, T. E., and Clayton, C. R., 1998, Interaction of sulphate-reducing bacteria with molybdenum dissolved from sputter-deposited molybdenum thin films and pure molybdenum powder, *J. Colloid. Interface Sci.*, 204, 237.
77. Percival, S. L., Knapp, J. S., Wales, D. S., and Edyvean, R. G. J., 1999, The effect of turbulent flow and surface roughness on biofilm formation in drinking water, *J. Ind. Microbiol. Biotechnol.*, 22, 152.
78. LeChevallier, M. W., Lowry, C. D., and Lee, R. G., 1990, Disinfecting biofilms in a model distribution system, *J. Am. Water Works Assoc.*, 82, 87.
79. Frateur, I., Deslouis, C., Kiene, L., Levi, Y., and Tribollet, B., 1999, Free chlorine consumption induced by cast iron corrosion in drinking water distribution systems, *Water Res.*, 33, 1781.
80. Kain, R. M., Tuthill, A. H., and Hoxie, E. C., 1984, Resistance of types 304 and 316 stainless steel to crevice corrosion in natural waters, *J. Mater. Energ. Syst.*, 5, 4.
81. Tverberg, J., Pinnow, K., and Redmerski, L., 1990, in *The Role of Steel*, Corrosion '90, National Association of Corrosion Engineers, Las Vegas.

11 Disinfection and Control of Biofilms in Potable Water

CONTENTS

11.1 INTRODUCTION

Disinfection is used in potable water treatment processes in order to reduce pathogens to an acceptable level and thus prevent public health concerns. However, scientific evidence is mounting, suggesting that exposure to chemical by-products formed during the disinfection process may be associated with adverse health effects. Reducing the amount of disinfectant or altering the disinfection process may decrease by-product formation; however, these practices may increase the potential for microbial contamination. Therefore, at present, it is necessary for research in the areas of potable water and disinfection to balance the health risks caused by exposure to microbial pathogens with the risks caused by exposure to disinfection by-products, specifically tri-halomethanes halomethanes.

In order for biocides to be effective in potable water they must

- Destroy all pathogens introduced into potable water within a certain time period at specified temperatures. This is particularly important as temperature and biocidal activity is loosely related, with biocidal properties reduced at lower temperatures owing to loss of enzyme activity.
- Be able to overcome fluctuations in composition, concentration, and conditions of waters which are to be treated.
- Not be toxic to humans or domestic animals nor unpalatable or otherwise objectionable in required concentrations.
- Be dispensable at reasonable cost, safe, and easy to store, transport, handle, and apply.
- Have their concentration in the treated water easily and quickly determined.
- Persist within disinfected water in a sufficient concentration to provide reasonable residual protection against possible recontamination from pathogens before use—the disappearance of residuals must be a warning that recontamination may have taken place.

Disinfection is an essential and final barrier against humans being exposed to all disease-causing pathogenic microbes, including viruses, bacteria, and protozoa parasites. Chlorine is an ideally suited disinfectant used in potable water. The reasons for this, as pointed out by Geldreich,[1] are owing to its availability and cost combined with its ease of handling and measurement, together with historical implications. However, in recent years, the finding that chlorination can lead to the formation of by-products that can be toxic or genotoxic to humans and animals has led to a quest for safer disinfectants. This is particularly important because the concentration of disinfectants is required in much higher levels needed to kill pathogenic microbes present within a biofilm when compared to their planktonic counterparts. This has led to the search for new disinfectants which could be both effective in potable water and, at the same time, cause destruction of microbes in biofilms. Options presently available as primary disinfectant alternatives to that of chlorine, include ozone, chlorine dioxide, and chloramines. Other useful ones include iodine, bromine, permanganate, hydrogen peroxide, ferrate, silver, UV light, ionising radiation, high pH, and the use of high temperature.

TABLE 11.1
The Advantages and Disadvantages of Disinfectants Used in Potable Water

Biocidal Treatment	Advantages	Disadvantages
Chlorine	Broad spectrum of activity	Produces toxic by-products
	Residual effect	Degradation of recalcitrant compounds to biodegradable products
	Generated on site	Reacts with extracellular polymers in biofilms
	Active in low concentrations	Low penetration characteristics in biofilms
	Destroys biofilm matrix	
Chloramines	Good penetration in biofilms	Less effective than chlorine to planktonic bacteria
	Reacts specifically with microorganisms	Resistance has been observed
	Low toxity by-products	Penetrates biofilms better than chlorine
Chlorine dioxide	Activity is not as pH dependent as chlorine	Explosive gas
	Effective in low concentrations	Safety problems
	Can be generated on site	Toxic by-products
Ultraviolet light	Efficient inactivation of bacteria and viruses	High doses required to inactivate cysts
	No production of known toxic by-products	No disinfectant residual in potable water
	No taste or odour problems	Difficulty in determining UV dose
	No need to store and handle toxic chemical	Biofilms may form on lamp surfaces
		Problems in the maintenance and cleaning of UV lamp
		Higher cost of UV disinfection than chlorination
Ozone	Similar effectivity as chlorine	Oxidises bromide
	Decomposes to oxygen	Reacts with organics and can form epoxides
	No residual	Degrades humic acids and makes them bioavailable
	Weakens biofilm matrix	Short half life
		Sensitive to water nutrients

Source: From *Wastewater Microbiology,* Bitton, G., Copyright © 1994. Reprinted by permission of Wiley-Liss, Inc., a subsidiary of John Wiley & Sons, Inc.

The effectiveness of a disinfectant is governed by the concentration of the disinfectant (C) which is measured in m/l per contact time (T) which is determined in minutes. These C/T values for all disinfectants are affected by a number of parameters including temperature, pH, disinfectant demand, cell aggregations, disinfectant mixing rates, and organics.

However, with the use of disinfection comes the formation of microbes which are resistant to disinfectants. Table 11.1 shows the disadvantages and advantages of disinfectants used in potable water.[7]

In both potable water and waste water, it is generally found that the organisms present can be classified under their resistance to disinfection. This is generally

$$\text{Coliforms} < \text{virus} < \text{protozoan cysts}$$

With respect to the main disinfectants used in water treatment and order of efficiency, it is generally found that the following pattern is seen with regard to coliform inactivation[1]

Ozone > chlorine dioxide > hypochlorous acid > hypochlorite ion > chloramines

Within laboratory studies in clean waters which have exerted no chlorine demand, it is possible to estimate the concentrations of disinfectant required to kill certain microbes. It is found that 3 to 100 times more chlorine is required to inactivate enteric viruses than is needed to kill coliform bacteria when external conditions such as temperature and pH are kept constant.

11.2 CONSIDERATIONS OF THE EFFECTS OF DISINFECTION ON BIOFILMS

A major decision regarding the choice of treatment for biofilms in potable water is related to whether its prevention or control of accumulation is desirable. Prevention requires disinfection of the incoming water, continuous flow of biocide at high concentrations, and/or treatment of the substratum which completely inhibits microbial adsorption. The extent to which any treatment can be applied depends on environmental, process, and economic consideration.

Generally, when considering the usage of biocides for the control of biofilm accumulation, a large number of factors have to be borne in mind. Commonly, the rate of cells' adsorbing to a substratum seems to be directly proportional to the concentration of cells in the bulk water. Therefore, by reducing the cell concentration in the bulk water, there will be a decrease in transport rate of cells. Ultimately, the reduced rate of cellular transport will reduce the rate of biofouling. Whilst filtering to remove bacteria is able to reduce cell numbers,[2] this can be a very expensive solution particularly when large volumes of water are used. Disinfection of the incoming water as in drinking water can be relatively effective in minimising biofilm accumulation. Nevertheless, the accumulation of an established biofilm (after a chlorine treatment) is owing primarily to growth processes and the contribution of the transport and attachment of cells.

The majority of the research looking at the efficiency of disinfectants on biofilms have been performed in laboratory-based studies. From these studies, it is found that microbial attachment to a surface results in decreased disinfection, particularly by chlorine.[3-5] Also, LeChevallier, Cawthon, and Lee[6] have shown that there is a decreased sensitivity to biocides when organisms are attached to a surface, with this effect greatly enhanced in older biofilms. This will have very important implications on any biofilm control regime unless appropriate monitoring is carried out.

11.3 CHLORINE

11.3.1 General Characteristics

Chlorine is the most commonly used biocide for controlling feacal coliforms, total coliforms, heterotrophic bacteria, and, also, biofouling within potable water systems.

TABLE 11.2
The Inactivation of Microorganisms by Chlorine: Ct Values (Temperature 5°C; pH 6.0)

Microorganism	Chlorine conc., mg/l	Inactivation Time (min)	Ct
E. coli	0.1	0.4	0.04
Poliovirus 1	1.0	1.7	1.7
G. lamblia cysts	1.0	50	50

Source: From *Wastewater Microbiology*, Bitton, G., Copyright © 1994. Reprinted by permission of Wiley-Liss, Inc., a subsidiary of John Wiley & Sons, Inc.

It is usually introduced into water as chlorine gas. Once introduced into water, it hydrolyses to[7]

$$\underset{\text{chlorine gas}}{Cl_2} + H_2O \rightarrow \underset{\text{hypochlorous acid}}{HOCl} + H^+ + Cl^-$$

$$HOCl \rightarrow \underset{\text{hypochlorite ion}}{H^+} + OCl^-$$

The proportion of HOCl and OCl^- are affected by the pH of water. Free chlorine consists of HOCl or OCl^-.

The reaction (depletion) of chlorine in the bulk water is generally referred to as the chlorine demand of water.[8] The chlorine demand is owed to soluble oxidizable inorganic compounds, soluble organic compounds, microbial cells, substratum, and particulate in the bulk water. It is now well documented that some materials and biofilms found in potable water have a chlorine demand which ultimately affects the efficiency of chlorination as a disinfectant. The inactivation of some microorganisms by chlorine is shown in Table 11.2.[7]

11.3.2 Mode of Action

Chlorine is known to have two types of effects on bacteria.[7] These are

1. Disruption of cell permeability—chlorine disrupts the integrity of the bacterial cell membrane leading to loss of cell permeability and, therefore, the leaking of proteins, DNA, and RNA.
2. Damage to nucleic acids and enzymes.

11.3.3 Effectiveness on Biofilms

The effectiveness of chlorine within potable water depends on its ability to inactivate sessile organisms and/or detach significant portions of the biofilm. Chlorine is seen as an effective microbial fouling control biocide because it has been shown to disrupt and loosen biofilms within potable water. Characklis[9] has found that when chlorine makes contact with a biofilm, a number of processes are known to occur. These include

1. Detachment of the biofilm.
2. Dissolution of biofilm components.
3. Disinfection.

However, there are a number of factors which are known to influence the rate and extent of the chlorine-biofilm reaction[9] and include

1. *Turbulent intensity* Transport of bulk water chlorine to the water biofilm interface is the first step in the chlorine–biofilm interaction. The transport rate increases with increasing bulk water concentration and turbulence.
2. *Chlorine concentration at the water biofilm interface* The transport of chlorine within the biofilm or deposit is a direct function of the chlorine concentration at the interface. Diffusion into the biofilm can be increased by increasing the chlorine concentration at the bulk water–biofilm interface. High chlorine concentrations for short durations are more effective than low concentrations for long periods assuming the same long term chlorine application rates for both, that is, the product of treatment concentration and duration.
3. *Composition of the fouling biofilm* The reaction of chlorine within the biofilm is dependant on the organic and inorganic composition of the biofilm as well as its thickness or mass. Disinfection in potable water systems is effective at low chlorine concentrations. However, in well developed biofilms, much of the material is extracellular and may compete effectively for available chlorine within the biofilm, thereby, reducing the chlorine available for killing cells. The substratum may also consume chlorine and thus may also compete for it.
4. *Fluid shear stress at the water–biofilm interface* Detachment and reentrainment of biofilm, primarily owing to fluid shear stress accompanies the reaction of biofilm with chlorine. Detachment of biofilm owing to chlorine treatment has been observed and the rate and extent of removal depend on the chlorine application and the shear stress at the bulk liquid interface.
5. *pH* The hypochlorous acid–hypochlorite ion equilibrium may be critical to performance effectiveness. OCl^- apparently favours detachment while HOCl enhances disinfection.

Chlorine is a useful biofouling control compound but in heavily contaminated waters is consumed in side reactions (chlorine demand reactions) and is rendered ineffective. Even copper–nickel alloys poses a significant chlorine demand. Therefore, water quality and the substratum composition are of the factors that must be considered in choosing a treatment program to minimise biofilm formation.

The rate at which chlorine is transported through the water phase to the biofilm depends on the concentration of chlorine in the bulk water and the intensity of the turbulence. The chlorine concentration in the bulk water is the net result of the chlorine addition minus the chlorine demand rate of the water. The chlorine concentration at the biofilm–water interface drives the reactions of chlorine within the biofilm. If the chlorine reacts rapidly with the biofilm, the concentration at the interface will be low and transport of chlorine to the interface may limit the rate of the overall process within the biofilm. By increasing the intensity of turbulence through increased flow rate, both the diffusion in the bulk water and the concentration at the biofilm–water interface will increase.

The transport of chlorine within the biofilm occurs primarily by molecular diffusion. Because the composition of the biofilm is some 96 to 99% water the diffusivity of chlorine in the biofilm is probably some large fraction of its diffusivity in water. In biofilms of higher density or in those containing microbial matter associated with inorganic scales, tubercles or sediment deposits diffusion of chlorine may be relatively low. Diffusion and the reaction of chlorine in a biofilm determine its penetration and, hence, its overall effectiveness.

Chlorine reacts with various organic and reduced inorganic components within the biofilm. It can disrupt cellular material (detachment) and inactivate cells (disinfection). In a mature, thick biofilm, significant amounts of chlorine may react with EPS, which are responsible for the physiological integrity of the biofilm. With regard to pH, chlorine has been found to be most effective at a pH of 6 to 6.5, a range at which hypochlorous acid predominates.

Much of the research performed which looks at the efficacy of disinfectants against biofilms has generally been done in laboratory-based studies. From a number of studies, it has been established that attachment of organisms to surfaces results in a decrease in disinfection by chlorine.[3-5]

It is accepted that chlorine is to some extent effective against bacteria in the planktonic phase but less effective against biofilms. However, the models available still suggest that there is a degree of unpredictability in this.[10] Other researchers[11] have shown that low concentrations of chlorine (20 µg per litre) used synergistically with low concentrations of copper (5 µg per litre) prevented growth of micro- and macrofouling organisms. LeChevallier, Cawthon, and Lee[12] showed a similar effect with 1 mg per litre of copper and 10 mg per litre of sodium chlorite exposed to *Klebsiella pneumonia* biofilms for 24 hours at 4°C.

11.4 CHLORAMINES

11.4.1 General Characteristics

Owing to the public health implications associated with the production of trihalomethanes from the chlorination process, chloramines have been proposed as the next best alternative. However, chloramines are not known to be very efficient biocides. In traditional chloramination processes, ammonia is added to water first followed by the addition of chlorine in the form of chlorine gas. The conversion rate of free chlorine to chloramines is, as with chlorination, dependant upon pH, temperature, and the ratio of chlorine to ammonia present.

TABLE 11.3
The Inactivation of Microorganisms by Chloramines: Ct Values

Microorganism	Water	Temp. °C	pH	Est. Ct
E. coli	BDF	5	9	113
Coliforms	Tap + 1%	20	6	8.5
Mycobacterium avium	BDF	17	7	ND
Mycobacterium intracellulare	BDF	17	7	ND
Poliovirus 1	BDF	5	9	1420
Hepatitis A	BDF	5	8	592

Note: BDF = buffered demand free water; ND = no data available.

Source: From *Wastewater Microbiology*, Bitton, G., Copyright © 1994. Reprinted by permission of Wiley-Liss, Inc., a subsidiary of John Wiley & Sons, Inc.

In potable water HOCl reacts with ammonia, resulting in the formation of inorganic chloramines

$$NH_3 + HOCl \rightarrow \underset{\text{monochloramine}}{NH_2Cl} + H_2O$$

$$NH_2Cl + HOCl \rightarrow \underset{\text{dichloramine}}{NHCl_2} + H_2O$$

$$NHCl_2 + HOCl \rightarrow \underset{\text{trichloramines}}{NCl_3} + H_2O$$

The proportion of these three forms of chloramines depends on the pH of the water with monochloramine predominate at pH greater than 8.5. Monochloramine and amine coexist between pH 4.5 and 8.5 and trichloramine at pH less than 4.5.

The use of chloramines has been shown to provide a long lasting, measurable disinfectant in potable water. Despite this, research has shown that monochloramines are definitely less effective disinfectants than free chlorine when compared at comparable low dose concentrations and short contact periods.

A major drawback of using chloramines in potable water, and for the control of biofilms, is that it is known to result in the formation of low concentrations of nitrites.[13] This may result in failures of potable water for nitrite standards, more so in the U.K. than the U.S. where standards for nitrite levels are less stringent. Although it is well known that nitrate levels have important implications on human health.

The inactivation of some microorganisms by chloramines is shown in Table 11.3.[7]

11.4.2 Mode of Action

The mechanism of action of monochloramine may account for its more effective penetration of bacterial biofilms than chlorine.[12] Monchloramine has been suggested to react rather specifically with nucleic acids, tryptophane and sulphur, containing amino acids but not with sugars such as ribose.[14]

11.4.3 Effectiveness on Biofilms

Chloramines have been shown to be very effective in suppressing biofilm development, particularly when water temperatures are above 15°C. They have been shown to be more effective than chlorine in reducing both sessile coliforms and also heterotrophic bacteria in potable water.[5,15] In one study, LeChevallier, Cawthon, and Lee[12] found that monochloramines are less effective than free chlorine against planktonic cells. The reverse was found when these disinfectants were exposed to sessile bacteria.

11.5 CHLORINE DIOXIDE

11.5.1 General Characteristics

Chlorine dioxide is a strong oxidant formed by a combination of chlorine and sodium chlorine which effectively inactivates bacteria and viruses over a broad pH range.[16] Until recently it was used primarily in the textile and pulp/paper industry as a speciality bleach and dye-stripping agent. It is often used as a primary disinfectant, inactivating bacteria and cysts. However, it is unable to maintain a residual effect long enough to be useful as a distribution system disinfectant. Despite this disadvantage, it does have advantages over that of chlorine in that it does not react with precursors to form THMs.

Chlorine dioxide is often commercially sold as stabilised chlorine dioxide which is actually sodium chlorite in a neutral solution. Sodium chlorite is much slower acting and less effective than chlorine and reacts with water to form two by-products. These are chlorite and, to some extent, chlorate. These compounds have been associated with the oxidation of heamoglobin[17] and, therefore, usage within potable water is restricted to a dosage of 1 mg per litre, which is not considered in many cases to be sufficient to provide good disinfection. Other problems associated with chlorine dioxide is in the development of taste and odours in some communities. However, chlorine dioxide can oxidize organic compounds such as iron and manganese and supress a variety of taste and odour problems.[18,19] Its effectiveness on a number of bacteria, including *E. coli* and *Salmonella*, has been noted and has found to be equal to and greater than free chlorine.[20]

Because chlorine dioxide is an explosive gas at concentrations above 10% in air, it is produced on site by mixing sodium chlorite with either inorganic (e.g., hydrochloric, phosphoric, and sulphuric acids) or organic acids (e.g., acetic, citric, and lactic acid) at or below pH 4.0. However, owing to the deadly nature of chlorine gas produced, handling is a primary limitation on the widespread use of chlorine dioxide.

Overall, the health concerns, tastes, odours, and relatively high cost, owing to generation of chlorine dioxide on-site and the concentrations that can be used in potable water to be effective, have tended to limit the uses of chlorine dioxide as a primary disinfectant for use in potable water. It has been noted in causing problems with the thyroid gland and inducing high serum cholesterol levels.[21] Despite this, many water companies have been successfully using chlorine dioxide as a primary disinfectant, particularly where the water is above pH 8.

TABLE 11.4
The Inactivation of Microorganisms by Chlorine Dioxide: Ct Values

Microorganism	Water	Temp. °C	pH	Time (min)	% Reduction	Ct
E. coli	BDF	5	7	0.6–1.8	99	0.48
Poliovirus 1	BDF	5	7	0.2–11.2	99	0.2–6.7
Hepatitis A	BDF	5	6	8.4	99	1.7

Note: BDF = buffered-demand free water; ND = no data available.

Source: From *Wastewater Microbiology,* Bitton, G., Copyright © 1994. Reprinted by permission of Wiley-Liss, Inc., a subsidiary of John Wiley & Sons, Inc.

The inactivation of a number of microorganisms by chlorine dioxide is shown in Table 11.4.[7]

11.5.2 Mode of Action

It is well documented that the mode of action of chlorine dioxide is primarily on the disruption of protein synthesis[22] and the outer membrane of gram-negative bacteria.[23] In viruses the mode of action has been identified as the protein coat[24] and the viral genome.[25]

11.5.3 Effectiveness on Biofilms

The Secretary of State's legal requirement is that the combined concentration of chlorine dioxide, chlorite, and chlorate should not exceed 0.5 mg per litre chlorine dioxide equivalent. In order to determine that this 0.5 mg per litre was actually capable of controlling the presence of biofilms and, in particular, *Legionella pneumophila*, a study was undertaken at the Building Services Research and Information Association (BSRIA). A full scale self-contained rig was built to represent an office's or residential building's water services for 50 people.[26]

The system was built in triplicate to allow thermal treatment to be compared with chlorine dioxide treatment in both hard and soft water. Sections of copper and glass reinforced plastic from the cold water storage tanks were removed from the system to allow analysis of biofouling before and during disinfection.

Results from the systems treated with chlorine dioxide demonstrated that control of *Legionella* within the biofilms took 20 days in the system using soft water and 30 days in the system using hard water.[27]

This may indicate that the scaling occurring owing to the hard water may have been acting as a protective barrier and preventing the chlorine dioxide from working as efficiently as it did in the soft water.

Other studies have shown that chlorine dioxide might kill all oocysts of *Cryptosporidium parvum* in slightly contaminated water[28] and may be particularly relevant if oocysts were enmeshed within a biofilm as demonstrated by Rogers and Keevil.[29]

11.6 OZONE

11.6.1 General Characteristics

Ozone is a pungent-smelling and unstable gas. As a result of its instability, it is generated at the point of use. An ozone-generating apparatus includes a discharge electrode. To reduce corrosion, air is passed through a drying process and then into the ozone generator. The generator consists of 2 plates or a wire and tube with an electric potential of 15,000 to 20,000 volts. The oxygen in the air is dissociated by the impact of electrons from the discharge electrode. The atomic oxygen combines with atmospheric oxygen to form ozone.

$$O + O_2 \rightarrow O_3$$

The resulting ozone–air mixture is then diffused into the water that is to be disinfected. The advantage of ozone is that it does not form THMs. As with chlorine dioxide, ozone will not persist in water decaying back to oxygen in minutes. Ozone is very effective in potable water to remove taste, odour, and colour because the compounds responsible for these effects are unsaturated organics. It is also used for the removal of iron and manganese. Ozone is seen as a very powerful disinfectant and is well known to be more effective in the inactivation of *Giardia* cysts than chlorine. Although ozone is not pH dependent, its biocidal activity decreases as the water temperature increases and so it may have limited effects in hot water systems. However, one major drawback of using ozone is the fact that the residuals are quickly dissipated. Its lifetime is usually less than 1 hour in most potable water systems.[30] Due to this, it is often necessary to use a secondary application of chlorine to provide disinfectant residual protection in potable water.

11.6.2 Mode of Action

Ozone has been reported to affect bacterial membrane permeability, enzyme kinetics, and also DNA.[31,32] It is also known to damage the nucleic acid core in viruses.[33]

11.6.3 Effectiveness on Biofilms

Ozone has been widely used in Europe and, in particular France, as a water disinfection in a number of water treatment plants[34] with a 1 to 2 mg per litre ozone dosage recommended for the treatment of domestic water. In terms of treating biofilms, ozone has been used in the treatment of *Legionella pneumophila* on water fittings in hospitals. Although the *L. pneumophila* was eradicated from the fittings, it was also removed from the control system which was ozone free. But this control system was subjected to other unforseen treatments such as flushing and unexpected chlorine concentration increases.[35] Carrying out disinfection trials within actual hospitals is very credible. However, unlike laboratory trials, there is the underlying problem that the system one is dealing with will have inherent mechanical nuisances and the system per se will not be under one's control.

11.7 ULTRAVIOLET LIGHT

11.7.1 General Characteristics

Ultraviolet (UV) disinfection was first used at the beginning of the century to treat water in Kentucky, but it was abandoned in favour of chlorination. Owing to technological improvements, this disinfection process is now regaining popularity, particularly in Europe.[36] UV disinfection systems use low pressure mercury lamps enclosed in quartz tubes. The tubes are immersed in flowing water in a tank and allow passage of UV radiation at germicidal wavelengths. However, transmission of UV by quartz decreases upon continuous use. Therefore, the quartz lamps must be regularly cleaned by mechanical, chemical, and ultrasonic cleaning methods. Teflon has been proposed as an alternative to quartz, but its transmission of UV radiation is lower than in the quartz systems.

UV performs well against bacteria and viruses. The major disadvantages for use in potable water are that it leaves no residual protection for the distribution system.

11.7.2 Mechanisms of Action

Studies with viruses have demonstrated that the initial site of UV damage is the viral genome, followed by structural damage to the virus coat.[37] UV radiation damages microbial DNA at a wavelength of approximately 260 nm. It causes thymine dimerization which blocks DNA replication and effectively inactivates microbes.

Microbial inactivation is proportional to the UV dose, which is expressed in microwatt per second per cm^2. The inactivation of microbes by UV radiation can be expressed by the following equation[38,39]

$$N/No = e^{-kpdt}$$

where No is equal to the initial number of microorganisms per ml; N is equal to the number of surviving microbes per ml; k is equal to the inactivation rate constant (μW per sect per cm^2); pd is equal to UV light intensity reaching the organisms (μW per cm^2); and t equals exposure to time in seconds.

The preceding equation is subject to several assumptions, one of which is that the log of the survival fraction should be linear with regard to time.[39] In environmental samples, however, the inactivation kinetics are not linear with time which may be owing to resistant organisms among the natural population and to differences in flow patterns.

The efficacy of UV disinfection depends on the type of microorganisms under consideration. In general, the resistance of microbes to UV follows the same pattern as with chemical disinfectants, which is as follows[40]

protozoan cysts > bacterial spores > viruses > vegetative bacteria

This trend is supported by Wolfe.[36] Table 11.5 gives an indication of the dosage required to inactivate a number of microorganisms associated with potable water. A virus such as hepatitis A requires a UV dose of 2700 μW per s/cm^2 for 1 log inactivation[36] but necessitates 20,000 mW per sec per cm^2 in order for a 3 log reduction to occur.[41]

TABLE 11.5
Dosage of UV Light Required to Inactivate Microorganisms

Microorganism	Dosage μW-s/cm^2
E. coli	3000
Legionella pneumophila	380
Poliovirus 1	5000
Giardia lamblia	63,000

Source: From *Wastewater Microbiology*, Bitton, G., Copyright © 1994. Reprinted by permission of Wiley-Liss, Inc., a subsidiary of John Wiley & Sons, Inc.

Many variables (e.g., suspended particles, chemical oxygen demand, colour) in potable and waste water affect UV transmission in water.[39,42] Several organic compounds (e.g., humic substances, phenolic compounds, lignin sulfonates from pulp and paper mill industry, ferric iron) interfere with UV transmission in water. Indicator bacteria are partially protected from the harmful UV radiation when embedded with particulate matter.[43-45] Suspended solids protect microorganisms only partially from the lethal effect of UV radiation. This is because suspended particles in water and waste water absorb only a portion of the UV light.[46]

11.7.3 Effectiveness on Biofilms

One major advantage of UV disinfection is that it is able to destroy microbial life in the water phase without the addition of anything to the water. However, when applied to the control of biofilms, this characteristic is also a disadvantage because UV disinfection leaves no residual. Hence, UV disinfection can control, for example, the incoming source water, thus in essence, supplying sterile water which will prevent biofilm formation. However, no distribution system or network will remain sterile following assembly and commissioning, so although UV may help to maintain the cleanliness of an already sterile system, additional chemical disinfectant such as chlorine or bromine are added post UV disinfection.

11.8 IONISATION

11.8.1 General Characteristics

The process of ionisation has been documented[47] over many years. It is based upon electrolysis in which ions undergo electron transfer at an electrode surface. In water services, the techniques are concerned with releasing silver and copper ions into the water by passing an electrical current between electrodes placed in running water. As the electrons pass between the anode (+ve) and cathode (–ve) one or two electrons are left behind on the anode surface. As the remaining electrons travel across to the cathode, they are driven away by the flow of the water into solution. These ions in solution represent charged atoms or groups of atoms where one or more electrons have been lost and the atom is no longer neutral but carries a charge. In the case of silver ions, these are designated Ag^+ where one electron is missing and the ions carry a single positive charge. In the case of the copper ions, these are designated Cu^+ or Cu^{2+} depending upon whether one or two electrons are missing. Ionisation units are used on location and, in general, consist of an electrode chamber and control unit. Typically, the chamber will contain silver–copper alloy electrodes of between 10 to 30% silver, dependent on manufacturer. The size and number of electrodes will be dependent upon the type of application according to the water volume, flow rate, and required microbial control. In a number of studies, the combination of these metals with halogenation has been shown to have applications in the disinfection of both recreational and potable water.[48]

11.8.2 Mode of Action

Ionisation has been used to control both waterborne bacteria[49] and viruses.[50,51] Copper ions kill bacteria by destroying cellular protein owing to the oxidation of sulphydryl groups of enzymes, thus interfering with respiration.[52,53] Silver ions also interfere with enzyme activity by binding to proteins whilst both ions bind to DNA molecules.[54] Advantages of ionisation are that a residual is maintained throughout the systems.

11.8.3 Effectiveness on Biofilms

The effectiveness of ionisation on waterborne organisms has been well proven.[48] The use of this technolgy against biofilms has been well studied, particularly against *L. pneumophila*.[27,54-57] However, there are problems with using this technology in the field where parameters cannot always be guaranteed to remain constant. In one study where ionisation was compared in soft and hard water, there were complications owing to the scaling up of the electrodes and pH of the water leading to failure to control *L. pneumophila*.[58,59] Although these problems were rectified, it demonstrates the inherent problems of controlling pathogens with automated disinfection processes that are susceptible to changes in water chemistry.

11.9 OTHER BIOCIDES USED IN POTABLE WATER

Potassium permanganate is often used in water supply treatment, particularly for the removal of taste, odour, and the metal ions, iron and manganese.[60-61] Potassium permanganate has also found uses in the disinfection of concrete, cement mortar lining, and asbestos cement surfaces. However, it has limited disinfection efficacy and is not as effective as the use of chlorine.[62,63] There is a possible benefit of using potassium permanganate as a peroxidant in the early stages of the treatment process because it is well known to reduce the growth of algae and slime bacteria.[1]

11.10 FUTURE METHODS IN THE CONTROL AND REMOVAL OF BIOFILMS

Brisou[64] suggested that methods have been developed that can release bacteria from surfaces. This release, using enzymes, can act on various levels of sessile bacteria

- Directly on microbial adhesions.
- On the structures of the media sensitive enzymes.
- On the bacterial polysaccharides produced during the colonisation of the interfaces, whether inert or live.
- On aggregates.

Brisou has shown that hydrolases can release bacteria from surfaces with exposure time with this enzyme, generally in 2 to 4 hours. These enzymes have been shown to free oligosaccharides and monosaccharides. If these could be applied to biofilms within potable water environments, it ultimately may enable identification of bacteria that have been unobtainable in the past and also release bacteria from the surface of potable water pipes enabling greater disinfection. Could this suggest an alternative in the future to the use of biocides? If this is both practical and feasible, more work is needed in this area to enable a better understanding of the processes involved and solutions to problems. However, the diverse range of bacteria found within the biofilm and the complexity of extracellular polymeric substances found make it a very hard and daunting task as an alternative short term solution. Care must be taken when an approach such as enzymes are used as an alternative to the use of biocides because one would not want to release the detachable biofilm straight into the consumer's tap without some other form of disinfection.

11.11 DISINFECTANT RESISTANT ORGANISMS

With the use and overuse of disinfectants in potable water comes the development of disinfectant resistant organisms. From the literature, it is generally found that different bacteria, viruses, and protozoa vary in their resistance to disinfectants. Parameters which are particularly relevant to this include pH, temperature, disinfectant concentration, and contact time because changing any one of these conditions will produce different rates of inactivation for the same organism. The major feature during the conduction of any disinfection regime to consider is that viruses and protozoa are more resistant to disinfection than enteric bacteria.

It is generally found that heterotrophic bacterial populations can be controlled to levels of 500 organisms per ml in many water supplies by the addition and maintenance of 0.3 mg per litre residual chlorine.[65] From this work, it was found that any further increases in the residual chlorine concentration did not result in any significant decreases in the heterotrophic bacterial densities. The reason for this was that organisms were being protected in sediment habitats and selective pressures operated inducing the growth of resistant organisms.

Within studies carried out on chlorine resistance of bacteria[4] present in potable water, the greatest resistant has been observed in gram-positive, spore forming bacilli, *Actinomycetes* and some *Micrococci*. These organisms were found to survive 2 min exposures to 10 mg per litre free chlorine. As a contrast, it was found that organisms most sensitive to chlorine contact were *Corynebacterium/Arthrobacter*, *Klebsiella*, *Pseudomonas/Alcaligenes*, *Flavobacterium/Moraxella*, *Acinetobacter*, and *Micrococcus*. It was also found in this study that these organisms were inactivated by 10 mg per litre or less of free chlorine.

11.12 SHORT TERM CONTROL OF BIOFILMS

In practice, the maintenance of an effective free chlorine residual concentration in a water system cannot be relied upon to prevent biofilm formation. A range of heterotrophs have been recovered from water containing concentrations of free chlorine of 0.1 to 0.5 mg per litre.[66] There are two reasons for this. In fast flowing pipe, there is always a thin layer of slower moving water (the viscous sublayer) just above the biofilm. Any disinfectants have to pass through this layer. Free chlorine is highly reactive, but it is not persistent so its ability to affect biofilms is reduced by the presence of the viscous sublayer. Chloramines are less reactive than free chlorine but more persistent and able to penetrate the laminar layer to a greater extent. Chlorine-based disinfectants are unable to penetrate deep into the biofilm because of its polymer gel structure preventing penetration.

The effect of chlorine to penetrate the biofilm is based on its oxidative properties. However, research has established that even high concentrations (10 mg per litre) are not sufficient to kill bacteria growing within a biofilm. Normally, however, a low concentration is required (3 to 5 mg per litre) of active chlorine to be sufficient for biofilm elimination. However, the disinfection effect of chlorine is affected by the age of the biofilm, the surface material, the encapsulation of microbes, and nutritive factors.

Mechanical cleaning is a decisive factor in combination with biocides in the elimination of biofilms from potable water pipelines. The forces achieved by high pressure water rinsing are an alternative to mechanical cleaning. However, induced breakage of biofilms is essential for the effective use of biocides. An extreme biofilm problem cannot be overcome using only shock treatment with biocides. The effect is temporary without a combined treatment using both biocides and mechanical cleaning. Otherwise, the microbes are back on the surface within a week after treatment. In an effective system, both types of treatment are essential.

Prevention methods within areas where biofouling is evident involve regular cleaning, but this does not prevent viable bacteria from recolonising the surface. Physical scouring or pigging helps control biofilm build up if combined with organic acids or alkalis. Other attempts include filtration devices which are quickly fouled, or UV radiation. Transmission of UV radiation, however, is decreased in turbid water and has poor penetration in microflocculations. Keevil and Mackernes[66] have suggested heating water systems to perhaps 70°C for 1 hour to control biofouling. This may have possible scalding effects if the correct safety procedures are not adhered to.

There is a growing awareness in the U.S. water industry that in some areas with persistent and widespread biofilm problems, it may be better to keep the biofilm undisturbed until mains cleaning can be arranged. This means avoiding flushing and minimising sudden changes in water flow rates.

Classical disinfectants as mentioned previously (e.g., chlorine or chloramines) are ineffective in controlling attached biomass. Thus it is necessary to control attached biomass using several techniques such as limiting biologically degradable organic carbon (BDOC) and the concentration of suspended bacterial cells in the water entering the distribution system. Studies using levels of active chlorine of 3 to 5 mg per litre is effective for eliminating biofilms.[67] However, the efficiency of any biocide as a means of disinfection will ultimately depend on BDOC levels of the water, nutrient levels within the biofilm, age of the biofilm, surface material, and amount of extracellular material present.[6]

It is found that polysaccharides which constitute the matrix of the biofilm can be penetrated easily suggesting that the age and characteristics of the microorganisms within the biofilm are not important for biocidal efficacy.[68] Whilst the effects of chloramine have been found not to be as effective as hypochlorite at reducing planktonic microbes, its effectiveness at penetrating the exopolymer matrix makes it a better candidate for usage in biofilm removal even when it was compared to hypochlorous acid, chlorine dioxide, and monochloramine on the same bacteria grown on a solid metallic surface. Monochloramine was the most effective for killing the sessile bacteria.[68]

Biofilm control in a distribution system is complicated and requires continuous action. The characteristics of the finished waters feeding the system have to be carefully controlled (low BDOC, low cell concentration). A secondary chlorination (i.e., chlorination of water already in the distribution system) is not a curative treatment, but an additional precaution for killing planktonic microorganisms; its efficiency is directly related to the previous organic matter reduction and a good hydraulic regime.

It generally is found that if a number of control measures are used for removing biofilms, reoccurrence of the problem begins again after only 1 week. [69]

11.13 LONG TERM CONTROL OF BIOFILMS IN POTABLE WATER

Long term biofilm growth seems to be difficult to stop, but there are several ways biofilms can be controlled. These include

- A reduction of nutrient levels in rivers and lakes would reduce the potential for biofilm growth in potable water systems as would nutrient reduction at the water treatment works. Without nutrients, biofilms are not able to thrive and mature. However, nutrient removal from aquatic environments or potable water involves expensive technologies and, therefore, will be a solution in years to come.
- Using materials in potable water systems which do not leach nutrients, thus reducing excessive biofilm formation. In the U.K., nonmetallic materials in contact with potable water must comply to BS 6920.
- Effective management of the hydraulics of distribution systems involving the avoidance of slow moving or stagnant pockets of water helps control biofilm maturation and the continuous presence of disinfectant residual has a suppressive effect.
- The treatment of source water according to internationally approved standards will destroy pathogenic organisms. The appearance of faecal organisms in potable water may be owing to their survival of disinfection because they may be protected within biofilms. The metabolic activity within a

biofilm may protect species sensitive to changes in pH or high oxygen tension of the circulating water.[66] Van der Wend, Characklis, and Smith[70] and Le Chevallier, Babcock, and Lee[69] show that in a potable water distribution system, even in the absence of chlorine, bacterial growth in the liquid phase is negligible. Essentially, only the bacteria in biofilm attached to the walls of the distribution pipeline are multiplying and, owing to shear loss, constitute one of the main causes of deterioration of microbiological quality of water distribution systems.

11.14 CONCLUSION

Biofilm control within potable water systems is very complicated and requires immediate action with respect to potential waterborne disease implications. The major control with respect to reducing biofilm accumulation is governed by the careful control of water BDOC levels and maintaining, but ultimately reducing, cell count levels. Post disinfection with the use of chlorine is by no means a curative measure, but a precautionary measure when biofilm growth and coliform aftergrowth is evident. Whilst biofilm development within potable water cannot be avoided, at present an emerging problem associated with them exists. It is related to the public health significance of growth as part of a biofilm where it is known that biocidal activity is greatly reduced. Also, the increasing isolation of bacteria resistant to present day disinfectant concentrations will only complicate the argument. Even if increased dosage of chlorine is the answer, it is well known that the development of secondary precursors will have a major long term effect on human health. It is not possible to fully assess the performance of disinfection on biofilms in potable water distribution systems owing to the constantly changing variables evident. Whilst these have been looked at in many pilot and laboratory-based experiments, they have led to a number of conflicting results and unanswered questions.

11.15 REFERENCES

1. Geldreich, E. E., 1996, *Microbial Quality of Water Supply in Distribution Systems*, Lewis, New York.
2. Percival, S. L., Knapp, J. S., Edyvean, R., and Wales, D. S., 1997, Biofilm development on 304 and 316 stainless steels in a potable water system, *J. Inst. Water Environ. Manage.*, 11, 289.
3 LeChevallier, M. W., Hassenauer, T. S., Camper, A. K., and McFeters, G. A., 1984, Disinfection of bacteria attached to granular activated carbon, *Appl. Environ. Microbiol.*, 48, 918.
4. Ridgeway, H. F. and Olson, B. H., 1982, Chlorine resistance patterns of bacteria from two drinking water distribution systems, *Appl. Environ. Mirobiol.*, 44, 972.
5. Berman, D., Rice, E. W., and Hoff, J. C., 1988, Inactivation of particle-associated coliforms by chlorine and monochloramines, *Appl. Environ. Microbiol.,* 55, 507.
6. LeChevallier, M. W., Cawthon, C. D., and Lee, R. G., 1988, Factors promoting survival of bacteria in chlorinated water supplies, *Appl. Environ. Microbiol.*, 54, 2492.
7. Bitton, G., 1994, *Wastewater Microbiology,* Wiley-Liss, New York.

8. Characklis, W. G., 1990, Microbial biofouling control, in *Biofilms,* Characklis, W. G., and Marshall, K. C., Eds., John Wiley & Sons, New York, 585.
9. Characklis, W. G., Trulear, M. G., Stathopoulos, N. A., and Chang, L. C., 1980, Oxidation and destruction of microbial films, in *Water Chlorination: Environmental Impact and Health,* Vol. 3, Jolley, R. L., Brungs, W. A., and Cumming, R. B., Eds., Ann Arbor Science, Ann Arbor, MI, 349.
10. Pirou, P., Dukan, S., and Jarrige, P. A., 1997, PICCOBIO: a new model for predicting bacterial growth in drinking water distribution systems, *Proc. Water Quality Techn. Conf.,* Boston, November 17-21, 1996, American Water Works Association, Denver, CO.
11. Knox-Holmes, B., 1993, Biofouling control with low levels of copper and chlorine, *Biofouling,* 7, 157.
12. LeChevallier, M. W., Cawthon, C. D., and Lee, R. G., 1988, Inactivation of biofilm bacteria, *Appl. Environ. Microbiol.,* 54, 2492.
13. O'Neill, J. G., Banks, J., and Jess, J. A., 1997, Biofilms in water mains—now under control, in *Biofilms: Community Interactions and Control,* Wimpenny, J., Handley, P., Gilbert, P., Lappin-Scott, H., and Jones, M., Eds., Boline, Cardiff.
14. Jacangelo, J. G. and Olivieri, V. P., 1985, Aspects of the mode of action of monochloramine, in *Water Chlorination, Chemistry, Environmental Impact and Health Effects*, Jolley, R. L., Brung, W. A., and Cumming, R. B., Eds., Lewis Publishers, Chelsea.
15. Neden, D. G., Jones, R. J., Smith, J. R., Kireyer, G. J., and Foust, G. W., 1992, Comparing chlorination and chloramines for controlling bacterial regrowth, *J. Am. Water Works Assoc.*, 84, 80.
16. Tanner, R. S., 1989, Comparative testing and evaluation of hard-surface disinfectants, *J. Ind. Microbiol.,* 4, 145.
17. Bull, R. J., 1982, Health effects of drinking water disinfectants and disinfectant by-products, *Environ. Sci. Technol.,* 16, 554A.
18. White, J. M., Labeda, D. P., LeChevallier, M. W., Owens, J. R., Jones, D. D., and Gauthier, J. L., 1986, Novel actinomycete isolated from bulking industrial sludge, *Appl. Environ. Microbiol.,* 52, 1324.
19. Montgomery, J. M., 1985, *Water Treatment Principles and Design,* John Wiley & Sons, New York.
20. Malpas, J. F., 1973, Disinfection of water using chlorine dioxide, *Water Treat. Exam,* 22, 209.
21. Condie, L. W., 1986, Toxicological problems associated with chorine dioxide, *J. Am Water Works Assoc.*, 78, 73.
22. Bernarde, M. A., Snow, N. B., Olivieri, V. P., and Davidson, B., 1967, Kinetics and mechanism of bactererial disinfections by chlorine dioxide, *Appl. Environ. Microbiol.*, 15, 257.
23. Berg, J. D., Roberts, P. V., and Matin, A., 1986, Effect of chlorine dioxide on selected membrane functions of *Escherichia coli, J. Appl. Bacteriol.,* 60, 213.
24. Olivieri, V.P., Dennis, W. H., Snead, M. C., Richfield, D. C., and Kruse, C. W., 1985, Mode of action of chlorine dioxide on selected viruses, in *Water Chlorination: Environmental Implications and Health Effects,* Vol. 5, Jolley, R. L., Brung, W. A., and Cumming, R. B., Eds., Ann Arbor Science, Ann Arbor, MI.
25. Taylor, G. R. and Butler, M., 1982, A comparison of the virucidal properties of chlorine, chlorine dioxide, bromine chloride and iodine, *J. Hyg.,* 89, 321.
26 Pavey, N. L. and Roper, M., 1988, *Chlorine Dioxide Water Treatment—for Hot and Cold Water Services*, TN2/98, Oakdale Printing, Surrey, 53.

27. Walker, J. T., Roberts, A. D. G., Lucas, V. J., Roper, M. M., and Brown, R., 1999, Quantitative assessment of biocide control of biofilms and legionella using total viable counts, fluorescent microscopy and image anaylsis, *Meth. Enzymol.,* 310, 629.
28 Peeters, J. E., Mazas, E. A., Masschelein, W. J., Martinez de Maturana, I. V., and Debacker, E., 1989, Effect of disinfection of drinking water with ozone or chlorine dioxide on survival of *Cryptosporidium parvum* oocysts, *Appl. Environ. Microbiol.,* 55, 1519.
29. Rogers, J. and Keevil, C. W., 1995, Survival of *Cryptosporidium parvum* in aquatic biofilm, in *Protozoal Parasites in Water,* Thompson, C. and Fricker, C., Eds., Royal Society of Chemistry, London.
30. Glaze, W. H., 1987, Drinking water treatment with ozone, *Environ. Sci. Technol.,* 21, 224.
31. Ishizari, K., Shinriki, N., and Ueda, T., 1984, Degradation of nucleic acids with ozone. V. Mechanism of action of ozone on deoxyribosenucleoside 5¢—monophosphates, *Chem. Pharm. Bull.,* 32, 3601.
32. Ishizari, K., Sawadaishi, K., Miura, K., and Shinriki, N., 1987, Effect of ozone on plasmid DNA of *Escherichia coli in situ, Water Res.,* 21, 823.
33. Roy, D., Wong, P. K. Y., Engelbrecht, R. S., and Chian, E. S. K., 1981, Mechanism of enteroviral inactivation by ozone, *Appl. Environ. Microbiol.,* 41, 718.
34. Miller, G. W., 1978, An assessment of ozone and chlorine dioxide technologies of treatment of municipal water supplies, Environmental Protection Series, EPA-600/2-78-147.
35 Edelstein, P. H., Whittaker, R. E., Kreiling, R. L., and Howell, C. L., 1982, Efficacy of ozone in eradication of *Legionella pneumophila* from hospital plumbing fixtures, *Appl. Environ. Microbiol.,* 44, 1330.
36. Wolfe, R. L., 1990, Ultraviolet disinfection of potable water, *Environ. Sci. Technol.,* 24, 768.
37. Rodgers, F. G., Hufton, P., Kurzawska, E., Molloy, C., and Morgan, S., 1985, Morphological response of human rotavirus to ultraviolet radiation, heat and disinfectants, *J. Med. Microbiol.,* 20, 123.
38. Luckiesh, M. and Holladay, L. L., 1944, Disinfecting water by means of germicidal lamps, *Gen. Electr. Rev.,* 47, 45.
39. Severin, B. F., 1980, Disinfection of municipal waste water effluents with ultraviolet light, *J. Water Pollut. Control Fed.,* 52, 2007.
40. Chang, J. C. H., Ossof, S. F., Lobe, D. C., Dorfman, M. H., Dumais, C. M., Qualls, R. G., and Johnson, J. D., 1985, UV inactivation of pathogenic and indicator microorganisms, *Appl. Environ. Microbiol.,* 49, 1361.
41. Baltigelli, D. A., Lobe, D., and Sobsey, M. D., 1993, Inactivation of hepatitis A virus and other enteric viruses in water by ultraviolet, *Water Sci. Technol.,* 27, 339.
42. Harris, G. D., Adams, V. D., Sorensen, D. L., and Dupont, R. R., 1987, The influence of photoreactivation and water quality on ultraviolet disinfection of secondary municipal wastewater, *J. Water Pollut. Control Fed.,* 59, 781.
43. Oliver, B. G. and Cosgrove, E. G., 1977, The disinfection of sewage treatment plant effluents using ultraviolet light, *Can. J. Chem. Eng.,* 53, 170.
44. Qualls, R. G., Flynn, M. P., and Johnson, J. D., 1983, The role of suspended particles in ultraviolet irradiation, *J. Water Pollut. Control Fed.,* 55, 1280.
44a. Qualls, R. G. and Johnson, J. D., 1983, Bioassay and dose measurement in U.V. disinfection, *Appl. Environ. Microbiol.,* 45, 872.
45. Qualls, R. G., Ossoff, S. F., Chang, J. C. H., Dorfman, M. H., Dumais, C. M., Lobe, D. C., and Johnson, J. D., 1985, Factors controlling sensitivity in ultraviolet disinfection of secondary effluents, *J. Water Pollut. Control Fed.,* 57, 1006.

46. Bitton, G., Henis, Y., and Lahav, N., 1972, Effect of several clay minerals and humic acid on the survival of *Klebsiella aerogenes* exposed to ultraviolet irradiation, *Appl. Environ. Microbiol.*, 23, 870.
47. Sykes, G., 1965, *The Halogens, in Disinfection and Sterilisation,* 1965, Chapman and Hall, London, 381.
48. Pyle, B. H., Broadaway, S. C., and McFeters, G. A., 1992, Efficacy of copper and silver ions with iodine in the inactivation of *Pseudomonas cepacia, J. Appl. Bacteriol.*, 72, 71.
49. Landeen, L. K., Yahya, M. T., and Gerba, C. P., 1989, Efficacy of copper and silver ions and reduced levels of free chlorine in inactivation of *Legionella pneumophila, Appl. Environ. Microbiol.,* 55, 3045.
50. Abad, F. X., Pinto, R. M., Diez, J. M., and Bosch, A., 1994, Disinfection of human enteric viruses in water by copper and silver in combination with low levels of chlorine, *Appl. Environ. Microbiol.,* 60, 2377.
51. Yahya, M. T., Straub, T. M., and Gerba, C. P., 1992, Inactivation of coliphage MS-2 and poliovirus by copper, silver, and chlorine, *Can. J. Microbiol.,* 38, 430.
52. Hugo, W. B. and Russel, A. D., 1982, Historial introducation, in *Principles and Practices of Disinfection, Preservation and Sterilisation,* Russel, A. D., Hugo, W. B., and Aycliffe, G. A. J., Eds., Blackwell, Oxford, 8.
53. Domek, M. J., LeChevallier, M. W., Cameron, S. C., and McFeters, G. A., 1984, Evidence for the role of copper in the injury process of coliform bacteria in drinking water, *Appl. Environ. Microbiol.,* 48, 289.
54. Yahya, M. T., Landeen, L. K., Messina, M. C., Kutz, S. M., Schulze, R., and Gerba, C. P., 1990, Disinfection of bacteria in water systems by using electrolytically generated copper: silver and reduced levels of free chlorine, *Can. J. Microbiol.,* 36, 109.
55. Mietzner, S., Schwille, R. C., Farley, A., Wald, E. R., Ge, J. H., States, S. J., Libert, T., Wadowsky, R. M., and Miuetzner, S., 1997, Efficacy of thermal treatment and copper-silver ionization for controlling *Legionella pneumophila* in high-volume hot water plumbing systems in hospitals, *Am. J. Infect. Contr.,* 25, 452.
56. Rohr, U., Senger, M., and Selenka, F., 1996, Effect of silver and copper ions on survival of *Legionella pneumophila* in tap water, *Zentralbl Hyg. Umweltmed.,* 198, 514.
57. Rogers, J., Dowsett, A. B., and Keevil, C. W., 1995, A paint incorporating silver to control mixed biofilms containing *Legionella pneumophila, J. Ind. Microbiol.,* 15, 377.
58. Pavey, N., 1966, *Ionisation Water Treatment—for Hot and Cold Water Services,* Bourne Press, Bracknell.
59. Walker, J. T., Ives, S., Morales, M., and West, A. A., 1999, Control and monitoring of biofouling using an avirulent *Legionella pneumophila* in a water system treated with silver and copper ions, in *Biofilms in Aquatic Systems,* Keevil, C. W., Godtree, A., Holt, D., and Dow, C., Eds., Society for Applied Microbiology, London, 131.
60. Cherry, A. K., 1962, Use of potassium permanganate in water treatment, *J. Am. Water Works Assoc.*, 54, 417.
60a. Pirou, P., Dukan, S., and Jarrige, P. A., 1997, PICCOBIO: a new model for predicting bacterial growth in drinking water distribution systems, *Proc. Water Quality Tech. Conf.,* Boston, Novenmber 17–21, 1996, American Water Works Association, Denver, CO.
61. Shull, K. E., 1962, Operating experiences at Philadelphia suburban treatment plants, *J. Am. Water Works Assoc.,* 54, 1232.
62. Cleasby J. L., Bauman, E. R., and Black, C. D., 1964, Effectiveness of potassium permanganate for disinfection, *J. Am. Water Works Assoc.,* 56, 466.

63. Buelow R. W., Taylor, R. H., Geldreich, E. E., Goodenkauf, A., Wilwerding, L., Holdren, F., Hutchinson, M., and Nelson, I. H., 1976, Disinfection of New Water Mains, *J. Am. Water Works Assoc.,* 68, 283.
64. Brisou, J. F., 1995, Biofilms — Methods for Enzymatic Release of Microorganisms, CRC Press, Boca Raton, FL.
65. Geldreich, E. E., Nash, H. D., Reasoner, D. J., and Taylor, R. H., 1972, The necessity of controlling bacterial populations in potable waters: community water supply, *J. Am. Water Works Assoc.,* 64, 596.
66. Keevil, C. W. and Mackerness, C. W., 1990, Biocide treatment of biofilms, *Int. Biodeterior.,* 26, 169.
67. LeChevallier, M. W., Lowry, C. H., and Lee, R. G., 1990, Disinfecting biofilm in a model distribution system, *J. Am. Water Works Assoc.,* 82, 85.
68. Nagy, L. A., Kelly, A. J., Thun, M. A., and Olson, B. H., 1982, Biofilm composition, formation and control in the Los Angeles aqueduct system, *Proc. Water Quality Tech. Conf.,* Nashville, TN.
69. LeChevallier, M. W., Babcock, T. M., and Lee, R. G., 1987, Examination and characterization of distribution system biofilms, *Appl. Environ. Microbiol.,* 53, 2714.
70. van der Wende, E., Characklis, W. G., and Smith, D. B., 1989, Biofilms and bacterial potable water quality, *Water Res.,* 23, 1313.

Index

C

D

E

F

G

H

I

K

L

M

N

O

P

R

S

X

Y

Z